GOD

AND THE ATOM

GOD
AND THE ATOM
VICTOR J. STENGER

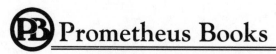

Prometheus Books

59 John Glenn Drive
Amherst, New York 14228–2119

Cover image © 2013 Media Bakery
Cover design by Grace M. Conti-Zilsberger

Inquiries should be addressed to
Prometheus Books
59 John Glenn Drive
Amherst, New York 14228–2119
VOICE: 716–691–0133 • FAX: 716–691–0137
WWW.PROMETHEUSBOOKS.COM

17 16 15 14 13   5 4 3 2 1

Library of Congress Cataloging-in-Publication Data forthcoming

Stenger, Victor J., 1935–
    God and the atom / by Victor J. Stenger.
        p. cm.
    Includes bibliographical references and index.
    ISBN 978–1–61614–753–2 (cloth : alk. paper)
    ISBN 978–1–61614–753–9 (ebook)

Printed in the United States of America on acid-free paper

CONTENTS

Preface   11

Acknowledgments   19

1. ANCIENT ATOMISM   21

Defining Atomism   21
Leucippus and Democritus   23
Atoms and Gods   25
Atoms and the Senses   26
Late Night with Lederman   27
Atomism in Ancient India   28
Epicurus   30
Differences with Democritus   33
Post-Epicurean Atomism   33
Lucretius   35
The Antiatomists   42

2. ATOMS LOST AND FOUND   47

Atomism in Early Christianity   47
Atomism in the Middle Ages   49
Poggio and Lucretius   51
Gassendi   56

3. ATOMISM AND THE SCIENTIFIC REVOLUTION 61

The New World of Science 61
Galilean Relativity 65
The *Principia* 67
Particle Mechanics 70
Mechanical Philosophy 72
Primary and Secondary Qualities 73
Other Atomists 75
More Antiatomists 76

4. THE CHEMICAL ATOM 79

From Alchemy to Chemistry 79
The Elements 83
The Chemical Atoms 84
The Chemical Opposition 86
The Philosophical Opposition 87

5. ATOMS REVEALED 91

Heat and Motion 91
The Heat Engine 93
Conservation of Energy and the First Law 95
The Mechanical Nature of Heat 97
Absolute Zero 99
The Second Law of Thermodynamics 100
Kinetic Theory 100
How Big Are Atoms? 102
Statistical Mechanics 104
The Arrow of Time 110

The Energetic Opposition   110
The Positivist Opposition   113
Evidence   117

6. LIGHT AND THE AETHER   119

The Nature of Light   119
The Aether   123
Fields   126
Electromagnetic Waves   130
The Demise of the Aether   131
Time and Space in Special Relativity   133
Defining Time and Space   137
Matter and Energy in Special Relativity   139
Invariance   140
Symmetry   141
The Source of Conservation Principles   142

7. INSIDE THE ATOM   145

Anomalies   145
Light Is Particles   147
The Rutherford Atom   150
The Bohr Atom and the Rise of Quantum Mechanics   152
Are Electrons Waves?   154
The New Quantum Mechanics   155
Spin   157
Dirac's Theory of the Electron   158
What Is the Wave Function?   159
The Heisenberg Uncertainty Principle   160
Building the Elements   162

8. INSIDE THE NUCLEUS 167

Nuclei 167
The Nuclear Forces 169
"Atomic" Energy 170
Nuclear Fusion 172
Nuclear Fission 174
Poisoning the Atmosphere 175
Nuclear Power 177
Liquid Fluoride Thorium Reactors 179

9. QUANTUM FIELDS 185

Physics in 1945 185
More Hydrogen Surprises 187
QED 188
Fields and Particles 194

10. THE RISE OF PARTICLE PHYSICS 197

Pion Exchange and the Strong Force 197
The Fermi Theory of the Weak Force 200
The Particle Explosion 202
New Conservation Principles 204
Broken Symmetries 206
"Nuclear Democracy" and *The Tao of Physics* 208

11. THE DREAMS THAT STUFF IS MADE OF

213

The Quarks   213
Particles of the Standard Model   216
Gauge Symmetry   219
Forces in the Standard Model   222
The Higgs Boson   226
Making and Detecting the Higgs   229
Hunting the Higgs   231
Higgs Confirmed!   234
Mass   234
Grand Unification   236
Supersymmetry   238

12. ATOMS AND THE COSMOS

241

After the Bang   241
Inflation   243
The Stuff of the Universe   246
What Is the Dark Matter?   248
Dark Energy   250
The Cosmological Constant Problem   252
Before the Bang   253
The Matter-Antimatter Puzzle   255
The Eternal Multiverse   256
Something about Nothing   257

13. SUMMARY AND CONCLUSIONS   261

They Had It (Mostly) Right   261
Matter   263
Materialism Deconstructed?   266
Field-Particle Unity   267
Wave-Particle Duality   269
Reduction and Emergence   270
The Role of Chance   272
The Cosmos   274
The Mind   275
No Higher Power   276

Notes   279

Bibliography   297

About the Author   309

Other Books by Victor J. Stenger   313

Index   317

PREFACE

*It is impossible for anyone to dispel his fear over
the most important matters, if he does not know
what is the nature of the universe but instead sus-
pects something that happens in myth. Therefore,
it is impossible to obtain unmitigated pleasure
without natural science.*

—Epicurus

Today every schoolchild knows that the world is made
of atoms. They have been taught that even a solid rock is
mostly empty space with tiny particles flitting about, occasionally
colliding with one another or sticking together to form more organized masses. Some adults may also remember this from school.

Since the atomic theory of matter was not fully confirmed until
the twentieth century, it is commonly thought that atoms are a
recent discovery of modern physics and chemistry. However, the
idea that everything is composed of infinitesimal, indivisible particles appeared in Greece 2,500 years ago and at about the same
time in India.

According to Aristotle (384–322 BCE), who disputed atomism,
Leucippus of Miletus (ca. fifth century BCE) invented the atomic
theory of matter. However, none of Leucippus's writing has survived, and his collaborator Democritus (ca. 460–ca. 370 BCE) is
credited with elaborating the theory. In the third century before
the Common Era, philosopher Epicurus (341–270 CE) built a whole
philosophy of life on the edifice of atomism.

Epicurus wrote extensively, but it was believed only a small

11

portion of his works had survived until a major work was recently discovered in the ruins of Herculaneum, which was destroyed by the eruption of Mount Vesuvius that also destroyed Pompeii. Unfortunately, this newly found work has not yet been translated to English. We would know little of Epicurean philosophy were it not for a magnificent 7,400-word poem in Latin hexameter called *De rerum natura* (*The Nature of Things*) written during the time of Julius Caesar by a Roman citizen named Titus Lucretius Carus (ca. 99–ca. 55 BCE).

The atomists proposed that not only is the stuff we see around us in the world made of atoms, but so is the soul or mind, which is therefore mortal and so dies with the rest of the body. There is no life after death. The gods exist, but their nature and role are unclear. They did not create the universe, which always existed and is unlimited in extent and contains multiple worlds. Nor did the gods create life. Rather, nature generated all kinds of "freaks" out of which only those adapted to the environment survived. Sounds a bit like natural selection, doesn't it?

In contrast to the Greeks, atomists in India regarded the soul itself as a separate, eternal atom. So Indian atomism was still dualistic while Greek atomism was monistic and atheistic at its core.

No one until the twentieth century had direct empirical evidence for atoms as particulate bodies. No one knows exactly how the original atomists arrived at their intuition. But observation must have played a role. No fact about the world has ever been discovered by pure thought alone. What was for millennia nevertheless a remarkable feat of human perception became the primary picture we have today of the nature of matter and the universe. But, from the beginning, the notion that everything is simply atoms and the void, with no divine creation or purpose, was in massive conflict with conventional thinking. Today this heresy is still vigorously opposed by some influential intellectual and religious elements in society. It is not that these opponents deny the overwhelming evidence for atoms. They simply reject the notion that they are all there is. This book will make the case that atoms and the void indeed are all there is.

Aristotle was opposed to atomism because, among other reasons, he believed that empty space was impossible. Aristotle's mentor, Plato, viewed matter as an illusion. Plato had an atomic idea of his own, where the elementary objects of reality are idealized, immaterial geometrical forms. Still, the polytheism of the ancient world was pretty flexible and tolerant of most beliefs. As long as the atomists paid lip service to gods of some sort, they could avoid serious trouble. It is only with monotheism that we began to see the forceful elimination of even the slightest deviations from the official state religious dogma.

Most authors who write on the subject insist that the ancient atomists were not atheists because they still believed in gods. Yes, they said they believed, but that was probably to avoid having to drink hemlock. The atomist gods play no role in the universe or in human lives, unlike theism as we understand it today. Atomism is atheism.

Atomism is also not deism. Unlike theism, deism is the belief in a creator god who does not involve itself with the universe or human lives. The universe of the atomists is eternal and uncreated. As we will see, the atomism of 2,500 years ago was essentially, in principle if not in detail, the model of the universe that modern science brings to us today.

The most influential philosophers of ancient Greece—Plato, Aristotle, and the Stoics—rejected material atomism. It conflicted too much with their ingrained beliefs in the world of gods and myths that had been passed down through the ages. With the rise of Christianity, which embraced Plato and Aristotle philosophically if not theologically, the works of Epicurus and Lucretius were suppressed during the thousand-year period from roughly the fifth to the fifteenth centuries. Through these years, known as the Dark Ages, the Roman Catholic Church dominated western Europe. It was only by sheer luck that a copy of *De rerum natura* survived to be rediscovered in a German monastery in 1417 CE. After another copy was taken to Florence where more copies were made, it became a sensation that played an important role in the nascent Renaissance and the scientific revolution that was soon to follow.

Although atoms would not be directly observed until the twentieth century, most physicists, including Galileo Galilei (1564–1642), adopted the atomic theory as the basic model for the primary structure of matter. While there was little speculation about the actual nature of atoms for lack of empirical data, the notion that point-like corpuscles move through space, colliding with one another and sticking together to form structures, was given a theoretical foundation by Isaac Newton (1642–1727) and those who followed. With Newtonian physics based on atomism, or at least on particle mechanics, the scientific revolution exploded on the world.

Not that there weren't doubters. Even by the late nineteenth century, after the atomic theory had proved enormously successful in explaining many observed phenomena involving gases and other fluids, the philosopher and physicist Ernst Mach (1836–1916) was prominent among many who refused to accept the reality of atoms. Mach held to the philosophy called *positivism*, in which only observable entities should be treated as real. Perhaps he would have changed his mind about atoms had he lived in the late twentieth century when he would have witnessed them as imaged on the screen of a device called the *scanning tunneling microscope*. Today we include many directly unobservable objects, such as quarks and black holes, in our theories.

Besides, as we will see, deciding on what is real and what is not real is no easy task. My basic position as an experimental physicist is that all we know about is what we observe with our senses and instruments. We describe these with models, sometimes called *theories*, but we haven't the faintest idea what is "really" out there. But, does it matter? All we need to concern ourselves with is what we observe. If whatever is really out there produces no observable effect, then why should we worry about it?

When the Reformation and Renaissance undermined Roman Church authority, new avenues of thought were opened up and science came into its own. Atomism—as a useful model—became an important part of the scientific revolution and eventually both Catholic and Protestant churchmen no longer saw it as the atheist threat it once surely was when articulated by Epicurus and Lucretius. Their theology was simply ignored by churchmen. After all, the ancients did not know Christ.

This book chronicles the empirical confirmation of atomism, from Leucippus and Democritus to Peter Higgs (and others), which reached its current form in the field where I spent my forty-year research career—that is, elementary particle physics. I will argue that the reduction of all we observe to the interaction of tiny bits of matter moving about mostly randomly in empty space is irreconcilable with the common belief that there must be something more to the universe we live in, that human thoughts and emotions cannot be simply the result of particles bouncing around. We will see how attempts to uncover evidence for immaterial ingredients or holistic forces in nature that cannot be reduced to the interactions of elementary particles have been a complete failure.

Before we begin our story, a few clarifications are needed. The term *atom* arises from the Greek word for *uncuttable*. The original notion of the ancient atomists was that the ultimate particles that make up matter cannot be further divided into more elementary parts. Today, based on history, we take a more cautionary, empirical approach and simply call the elementary particles of matter in our models "conceptually indivisible." So, for example, in the current so-called standard model of elementary particles and forces, electrons and quarks are uncuttable; but we can never say for sure they will always remain that way as science advances.

In the nineteenth century, the elements of the chemical periodic table seemed to be uncuttable and, indeed, are still called "atoms." Chemists were unable to subdivide the elements simply because the energies involved in chemical reactions, produced by Bunsen burners and electric sparks, are too low. Once much higher energies became available with nuclear radiation and particle accelerators, it was discovered that the chemical elements were not elementary after all but that each element is composed of a tiny nucleus surrounded by a cloud of electrons.

It did not end there. Nuclei were found to be composed of protons and neutrons and these, in turn, were discovered to be made of more elementary particles we identify as *quarks*. At this writing, the set of elementary particles in the standard model includes quarks, electrons, and other particles such as neutrinos and photons for which no substructures have yet been identified in experiments. This model has been in existence since the 1970s, agreeing with all observations, and only now are experiments reaching sufficiently high energy where further substructure might be revealed. This book is being written just as the final gap in the standard model, the Higgs boson, seems to have been filled at the Large Hadron Collider (LHC) in Geneva, Switzerland. No one is stopping there. More data from the collider will surely, even necessarily, point us in the direction of new physics beyond the standard model.

It is felt by most physicists that ultimately we will have to reach the point where the ultimate uncuttable constituents of matter will be established. I will generally call them *elementary particles* rather than "atoms" and, to further avoid confusion, the structures that constitute the chemical elements will be referred to as *chemical atoms*, unless the distinction is clear from context.

The ancient atomists introduced the notion that atoms move around in otherwise empty space—a vacuum or a void. As mentioned, Aristotle attempted to prove that such a void was impossible, but when Evangelista Torricelli (1608–1647) and other seventeenth-century scientists began producing vacuums in the laboratory, Aristotle's views fell out of favor. Of course, these laboratory vacuums, then and now, are hardly empty space. But at least the notion was established that a chamber empty of particles is conceivable.

Today we often hear it said that, according to quantum mechanics, we can never have completely empty space, as particle-antiparticle pairs flit in and out of existence. While this is true, at any given instant a volume will contain these particle pairs with empty space in between. The basic atomic model remains part of quantum physics. The matter we observe on all scales is mostly empty space with tiny particles mostly randomly moving about constituting the visible universe and perhaps its invisible parts as well.

Yet another clarification is needed because of the use of "particles" in the preceding paragraphs. It remains possible that in some future, successful theory, the ultimate constituents or atoms of matter may not be treated as point-like (zero-dimensional) particles but strings (one-dimensional) or multidimensional "branes" (from "membranes"). Even if these models ultimately succeed (they haven't so far), the elementary structures will be so small that they will remain particulate in the eyes and instruments of experimenters for the foreseeable future. For my purposes, I have no need to bring in these speculations and will stick to what is already well established.

Some confusion may also arise when we discuss the issue of reductionism. I will claim that the atomic model exemplifies the notion that we can reduce everything to its parts. Despite desperate opposition from those wedded to holistic philosophies, reductionism has triumphed.

However, you might wonder, if an atom were "uncuttable," then that would seem to mean that it is irreducible. If that is the case, then how can atomism be reducible?

The reducibility of the atomic model refers to the fact that the observations we make about matter, such as the wetness of water or the color of copper, and perhaps even human intelligence, can be reduced to the motions and interactions of elementary particles that themselves do not possess such properties. The anti-reductionists have always objected that this is impossible. We will give examples showing that it does indeed happen. And, as I said, nothing is stopping us from considering the current elementary particles as ultimately reducible to even smaller parts. This is physics, not philosophy. What matters is data, not words.

Finally, we will find that the expedient of describing an observed phenomenon in terms of the behavior of constituent particles of the material bodies involved not only greatly simplifies the understanding of these phenomena but also removes much of the mystery that confounds much of modern life in the physical world.

A note to the reader: This book starts out mostly historical and philosophical, but as it progresses chronologically, it becomes increasingly scientific. Some of the latter material is somewhat technical with a few equations at the level of high-school algebra, but it still should be accessible to nonscientists who have at least some familiarity with the subjects from reading popular books and articles. I feel that a minimum amount of technical detail is necessary to establish the validity of my thesis, that modern science has fully confirmed the model of the world first proposed 2,500 years ago.

ACKNOWLEDGMENTS

I am deeply grateful for the invaluable comments, suggestions, and corrections I received from historian Richard Carrier, philosopher Tom Clark, and the following members of the Internet discussion list avoid-L that tirelessly provides me with feedback on my writing: Martin Bier, Lawrence Crowell, Keith Douglas, Yonatan Fishman, John Mazetier, Don McGee, Brent Meeker, Kerry Regier, Anne O'Reilly, Christopher Savage, and Bob Zannelli. I would also like to thank the staff of the public library in Louisville, Colorado, for the efficiency with which they provide me with books from other libraries throughout the state.

1

ANCIENT ATOMISM

The universe consists of bodies and void: that bodies exist, perception itself in all men bears witness; it is through the senses that we must by necessity form a judgment about the imperceptible by means of reason.

—**Epicurus**[1]

DEFINING ATOMISM

In his exhaustive study *Atomism and Its Critics: Problem Areas Associated with the Development of the Atomic Theory of Matter from Democritus to Newton,* philosopher Andrew Pyle lists what he defines as the *ideal* central claims of atomism:

1. A commitment to *indivisibles*, particles of matter either conceptually indivisible (i.e., such that one cannot conceive of their division) or physically unsplittable.
2. Belief in the existence of *vacuum* or "Non-Being," purely empty space in which the atoms are free to move.
3. *Reductionism*: explanation of the coming-to-be, ceasing-to-be and qualitative alternation of sensible bodies in terms of the local motion of atoms which lack many (most) of the sensible properties of those bodies.
4. *Mechanism*. This is the thesis about the nature of physical

agency: it claims in effect that no body is ever moved except by an external impulse from another body.[2]

The book is Pyle's doctoral dissertation at the University of Bristol. Its chapters deal with each of the above four issues over three periods: classical antiquity (ca. 500 BCE–500 CE), the Middle Ages and Renaissance (ca. 500–1600), and the seventeenth century.

Pyle tells us that many Renaissance thinkers accepted (3) but not (4), "insisting that the movements of minute bodies that constitute the generation and alteration of sensible bodies are guided by purposive, spiritual agencies of some kind."[3]

In this book, I will carry this discussion on to the present day, showing how atomism—including item (4)—constitutes our best description of the observations we make of the world. We will see how material atomism was resisted by many of the greatest thinkers of all time, from Aristotle and Plato to Augustine and Aquinas, and then on to the present day, where we find it under attack by those who refuse to believe that matter and natural forces are all there is to observable reality.

The reductionism in item (3) that forms a part of the doctrine of atomism is highly unpopular. Time and again, we hear from scientists, philosophers, theologians, spiritualist gurus, and laypeople that "the whole is greater than the sum of its parts." We will see that while this statement is technically true, it is far less profound than its proponents claim. As Alex Rosenberg notes in *Darwinian Reductionism; or, How to Stop Worrying and Love Molecular Biology*:

> [The] whole has properties which no individual part has: the property of wetness that water has is not a property that any H_2O molecule has. But this is no reason to deny the reducibility of wetness to properties of H_2O molecules. Wetness is identical to the weak adhesion of these molecules to one another, owing to the polar bonds between the molecules; and these bonds are the result of the distribution of electrons in the outer shells of the constituent atoms of the H_2O molecules.[4]

In other words, the whole still results from the action of its parts. Try as they might, the anti-reductionists have been unable to find any evidence to support their distaste for atomism. No special holistic forces have been shown to come into play with the aggregation of large numbers of parts; just new properties develop or, in the common parlance of today, "emerge" from the interaction of the parts.[5]

LEUCIPPUS AND DEMOCRITUS

Leucippus, who lived in Miletus (or possibly Elea or Abdera) in Ionia in the early fifth century before the Common Era, is usually credited with inventing the theory of atoms and the void, at least the Greek version that has come down to us. Little is known about him, and none of his writing has survived. More is known about Democritus of Abdera, who is thought to have been Leucippus's student, or at least a much younger colleague, and he appears in the anecdotes of many ancient texts. He is reported to have produced a large number of works on many subjects, but these have only survived in secondhand reports.[6]

The atomic theory of Leucippus and Democritus can be characterized by the simple phrase "atoms and the void." Everything is made of atoms, even gods and the soul. While today we think of atoms as moving around in empty space, or "nothing," Leucippus and Democritus did not regard the void as "nothing." It is just as much a part of reality as "something." A single, infinite entity exists in reality. That reality breaks up into an infinite number of infinitesimal parts—atoms and the void. In this way, the concept of a unique elementary substance is retained while accounting for diversity and the potential for change.

According to these early atomists, the atoms themselves are hard, incompressible, and indivisible. They have no parts. Although lacking any substructure, they have different gross geometrical characteristics: size, weight, shape, order, and position. It is not clear whether the property of weight was introduced by Democritus or later by Epicurus.

They argued that the motions of atoms are endless and largely random, with a tendency to move toward some point such as the center of Earth. Everything happens by either chance or necessity. When they collide, atoms either recoil from one another or coalesce according to their various shapes. For example, they may have hooks that enable them to grab onto one another. When they join to form new, compound identities, individual atoms still retain their original identities.

In the third century of the Common Era, the historian Diogenes Laertius summarized Democritus's atomic model as follows:

> The principles of all things are atoms and the void, and everything else exists only by convention. The worlds are unlimited and subject to generation and corruption. Nothing could come to be from nonbeing, and nothing could return by corruption to nonbeing. Atoms are unlimited in size and number, and are the seat of a vortex motion in the universe, which results in the creation of all compounds: fire, water, air, and earth, which are simply organizations of certain atoms, themselves resistant to change and alteration by virtue of their hardness. The sun and the moon are composed of such particles, smooth and round, as is the soul, which is the same thing as the intellect.[7]

Note that fire, water, air, and earth are not the basic elements in this scheme, as they were held to be in almost all other ancient natural philosophy, either as the individual primal stuff, or in combination. Even in the Middle Ages, alchemy was based on the principle that you could combine these elements to make other compounds. In particular, by adding fire to earth (stone), you could make gold. That never succeeded, but in the twentieth century, physicists were able to combine the nuclei of baser elements, what I will call *chemical atoms*, to make gold. Unfortunately, they couldn't make enough of the precious metal to pay for the experiment, much less generate riches.

The vortex mentioned above is suggestive of the swirling nebula of interstellar matter that contracted under gravity to produce our solar system. Atoms tended to move in that direction.

We will have more to say later about whether nothing can come from nonbeing. The basic point here is that atoms, as described by Democritus, always existed and are indestructible, while the compounds they form, including the soul, can come and go.

ATOMS AND GODS

From the beginning, atomism has been an anathema to religious belief. According to philosopher David Sedley:

> Atomism [is] the first Presocratic philosophy to eliminate intelligent causation at the primary level. Instead of making intelligence either an irreducible feature of matter, or, with Anaxagoras, a discrete power acting upon matter, early Greek atomism treats atoms and void alone as the primary realities, and relegates intelligence to a secondary status: intelligence, along with color, flavor, and innumerable other attributes, is among the properties that supervene on complex structures of atoms and the void.[8]

Democritus was a contemporary of Socrates, so he was well aware of the dangers of public impiety. However, he assigned a limited role to the gods. They did not intervene in a world governed by natural laws. Furthermore, since the soul is made of atoms, it is not immortal and humans can never become gods.

ATOMS AND THE SENSES

In *The Presocratic Philosophers*, philosophers Geoffrey Kirk, John Raven, and Malcolm Schofield explain how Democritus viewed the senses:

> Democritus sometimes does away with what appears to the senses, and says that none of these appears according to truth but only according to opinion: the truth in real things is that there are atoms and the void. "By convention sweet," he says, "by convention bitter, by convention hot, by convention cold, by convention color; but in reality atoms and the void."[9]

Democritus regarded observations such as color and taste as conventions and, thus, not real. Only atoms and the void are real, and these he could not see.

Although he quite correctly questioned the reliability of the senses, Democritus hardly ignored them. In fact, he proposed an atomic mechanism to explain how the senses operate. According to Democritus, visual perception results from atomic emanations from the body colliding with atoms in the eye. This is essentially as we understand sight today. Particles, or atoms, of light called *photons* are emitted from bodies. If the body is the sun or a lamp, the photons are energetic enough to excite electrons in the chemical atoms of the eye and produce a signal to the brain. Colder bodies, such as you and I, also emit photons, but these are in the lower energy infrared region of the spectrum, where our eyes are insensitive. In order to see human bodies directly by their emissions, our eyes must be aided by special night-vision goggles that detect infrared light. Since most of the objects surrounding us are not as hot as the sun, we see them by means of the higher energy photons from the sun or lamps as they scatter off the object and into our eyes.

The senses of sound and smell were also correctly viewed in

ancient atomism as the emanation and absorption of atoms. The sense of touch was also accurately described as a collision of the atoms of the hand, for example, with the atoms of the object being touched.

LATE NIGHT WITH LEDERMAN

In his highly entertaining book, *The God Particle*, to which I will refer several times in this book (which I wish I could make half as entertaining), Nobel-laureate physicist and former Fermilab director Leon Lederman imagines a dream in which he takes Democritus on a tour of the accelerator.[10] Here are two related excerpts:

> *Lederman:* How did you imagine the *indivisibility* of atoms?
> *Democritus:* It took place in the mind. Imagine a knife of polished bronze. We ask our servant to spend an entire day honing the edge until it can sever a blade of grass held at its distant end. Finally satisfied, I begin to act. I take a piece of cheese . . .
> *Lederman:* Feta?
> *Democritus:* Of course. Then I cut the cheese in two with the knife. Then again and again, until I have a speck of cheese too small to hold. Now, I think that if I myself were much smaller, the speck would appear large to me, and I could hold it, and with my knife honed even sharper, cut it again and again. Now I must again, in my mind, reduce myself to the size of a pimple on an ant's nose. I continue cutting the cheese. If I repeat the process enough, do you know what the result will be?
> *Lederman:* Sure, a feta-compli.

And a little later in the dream . . .

Lederman: Today we can almost define a good scientist by how skeptical he is of the establishment.

Democritus: By Zeus, this is good news. What do you pay mature scientists who don't do windows or experiments?

Lederman: Obviously you're applying for a job as a theorist. I don't hire too many of those, though the hours are good. Theorists never schedule meetings on Wednesday because it kills two weekends. Besides, you're not as anti-experiment as you make yourself out to be. Whether you like it or not, you did conduct experiments.

Democritus: I did?

Lederman: Sure. Your knife. It was a mind experiment, but an experiment nonetheless. By cutting a piece of cheese in your mind over and over again, you reached your theory of the atom.

Democritus: Yes, but it was all in the mind. Pure reason.

Lederman: What if I could show you that knife?

Democritus: What are you talking about?

Lederman: What if I could show you the knife that can cut matter forever, until it finally cut off an a-tom?

Democritus: You found a knife that can cut off an atom? In *this* town?

Lederman: [*nodding*] We're sitting on the main nerve right now.

Democritus: This laboratory is your knife?

Lederman: The particle accelerator.

ATOMISM IN ANCIENT INDIA

A form of atomism can also be found in ancient India. Although more closely tied to religion in very important ways than Greek atomism, enough similarities exist to lead one to suspect some contact between the two distant cultures.

According to historian Bernard Pullman, six major philosophical systems emerged from Hindu Brahmanism. Of these, the Nyaya-Vaisheshika movement was the strongest defender of atomism. A doctrine of atoms can be found in the Vaisheshika sutra, written by Kanada in the first century BCE. Other Hindu schools were receptive to the idea, while Vedanta was opposed.[11]

Kanada's atoms, like those of Democritus, were eternal, indestructible, innumerable, and without parts. However, they included the classical four elements—fire, air, water, and earth—along with aether, space, time, and two kinds of souls. Gods and individual humans contain eternal, omniscient souls, while another form of soul called *manas* is an organ of thought that acts as the link between the god-human soul and external objects.[12]

The Buddhist school of Hinayana held to an atomic doctrine similar to Nyaya-Vaisheshika. It also regarded the four elements as atoms, although it considered soul and conscience to be outside the realm of atoms. Jainism, on the other hand, seemed to hold views similar to the Greek atomists in not regarding the four elements as atoms.

All these philosophies viewed the soul as eternal and incorruptible, while the Greek atomists said the soul was a composite of atoms and thus disintegrates upon death along with the rest of the body.[13] In short, the atomism that came out of ancient India maintained the dualism of matter and mind/soul/spirit that is present in most religions while Greek atomism, although accepting the existence of remote gods uninterested in humanity, was distinctly atheistic—at least as the term is used today. In this book, I will take theism to be a belief in a superhuman god or superhuman gods existing outside the material world but still very much active in the operation of the universe and in human affairs. Atheism is nonbelief in such a god or gods, or in any kind of external, immaterial supernatural force. The gods of Greek atomism were not supernatural. They were made of atoms, too, and did not remotely resemble the superhuman gods of the *Iliad* and the *Odyssey* or the gods of India. Because of its duality, Indian atomism should not be considered comparable to that of the Greeks.

EPICURUS

Epicurus was born on the island of Samos in 341 BCE to poor Athenian parents. After compulsory military service, he studied philosophy for ten years in Teos in Ionia under Nausiphanes, from whom he learned the atomism of Democritus. Epicurus never acknowledged any contribution from Leucippus.

The Epicurean movement began after Epicurus moved to Colophon in Ionia. Gathering disciples along the way, he briefly taught on the island of Lesbos and in Lampsacus in Ionia, where he gained many more disciples and financial support. In 306 BCE, he settled in Athens where he held meetings in the garden of his house. Because of this, his movement became known as "The Garden."

Students of The Garden, which (scandalously) included both sexes, were expected to live a simple life of quiet study and to withdraw from politics. Epicurus developed a unique world-system based on atomism that repudiated most of the teachings of Aristotle, Plato, and all the other philosophical schools of his time and place: Our earthly life is all there is. The punishing and vengeful gods of myth do not exist. What gods do exist have no concern for humans because they live outside our world in a state of perfect happiness. Humans decide proper conduct by reasoning about the best actions to pursue. Justice is defined as dealing with others for mutual advantage and can change as circumstances change.

Epicurus died at age seventy in 271 BCE. He left behind over three hundred books, including *On Atoms and Void*. Unfortunately, only a small portion of his work remains, mostly fragmentary. These fragments can be found in *The Essential Epicurus: Letters, Principal Doctrines, Vatican Sayings, and Fragments* by Eugene O'Connor, which has been my primary reference for this section.[14] Pieces of Epicurus's masterwork *On Nature* were recovered from the ashes of Herculaneum, which was destroyed in the eruption of Mount Vesuvius that destroyed Pompeii in 79 CE. Unfortunately, so far these have not been widely available.

As you can imagine, Epicurus had many enemies and *Epicureanism* has long been wrongfully associated with a debauched life in the hedonistic pursuit of pleasure. In fact, while Epicurus regarded pleasure as an ultimate good, he was mainly concerned with avoidance of fear and pain by limiting desires and living modestly. Tranquility and freedom from fear and pain resulted from living a simple life of friendship, learning, and temperance.

In *Lives of the Eminent Philosophers,* Diogenes Laertius says that Epicurus had legions of followers and was honored in Athens with bronze statues. Early Christians approved of Epicurus's denouncing of pagan superstition, but his teachings were ultimately suppressed in medieval Christendom. In his epic fourteenth-century poem *Inferno,* Dante Alighieri (ca. 1265–1321) consigned Epicurus to the sixth circle of hell for denying the immortality of the soul (canto 10.13–15). Epicurus is the only ancient philosopher in Dante's hell.

As we will see later, the Renaissance saw a revival in interest in Epicurus, especially after the discovery of Lucretius's *De rerum natura.* In 1647 Pierre Gassendi published *Eight Books on the Life and Manners of Epicurus* that had great success, especially in England where it influenced Thomas Hobbes, John Locke, and other important figures.

De rerum natura is the main source we have today for Epicurus's teachings on atomism. However, before we get to that, let us look at some of his teachings by quoting directly from Epicurus's surviving works as translated by O'Connor. (The page numbers from that reference are given in parentheses.)

Quotations from Epicurus on atomism come from "Letter to Herodotus" (a friend of Epicurus's, not the ancient historian).

The universe consists of bodies and void: that bodies exist, perception itself in all men bears witness; it is through the senses that we must by necessity form a judgment about the imperceptible by means of reason. (21)

The universe is without limit. . . . Also, the universe is boundless both in the number of bodies and the magnitude of the void. . . . Moreover, there are infinite worlds, both like and unlike this one. (22–23)

Atoms exhibit none of the qualities belonging to visible things except shape, mass, and size, and what is necessarily related to shape. For every quality changes; but the atoms do not change, since, in the dissolution of compound substances, there must remain something solid and indestructible. (27)

The atoms must possess equal velocity whenever they move through the void, with nothing coming into collision with them. (30)

Let us also look at some of the words of Epicurus concerning religion. In his "Letter to Menoeceus," he talks about an immortal god but tells us not to apply anything to him "foreign to his immortality or out of keeping with his blessedness." He says that the assertions made about gods by the many are "grounded not in experience but in false assumptions" that the gods are responsible for good and evil. (62)
Epicurus is most eloquent when he speaks of death:

Grow accustomed to the belief that death is nothing to us, since every good and evil lie in sensation. However, death is the deprivation of sensation. Therefore, correct understanding that death is nothing to us makes a mortal life enjoyable, not by adding an endless span of time but by taking away the longing for immortality. For there is nothing dreadful for the man who has truly comprehended that there is nothing terrible in not living. Therefore, foolish is the man who says he fears death, not because it will cause pain when it arrives but because anticipation of it is painful. What is no trouble when it arrives is an idle worry in anticipation. Death, therefore—the most dreadful of evils—is nothing to us, since while we exist death is not present, and whenever death is present, we do not exist. It is nothing to the living or the dead, since it does not exist for the living and the dead no longer are. (63)

DIFFERENCES WITH DEMOCRITUS

According to the peer-reviewed *Internet Encyclopedia of Philosophy*, Epicurus modified the teachings of Democritus in three important ways:[15]

1. *Weight*. Aristotle had criticized Democritus for not explaining why atoms moved in the first place. Epicurus said that atoms had a natural motion—that is, "downward"—and proposes weight as the atomic property that accounts for this motion.
2. *The Swerve*. If atoms all just moved downward, they would never collide. So Epicurus added the notion that at random times they swerve to the side.
3. *Sensible Qualities*. As we have seen, Democritus said only invisible atoms and the void exist and sensible qualities such as sweetness are simply conventions. This led him to be pessimistic about our ability to obtain any knowledge about the world through our senses. Epicurus agreed that the properties of macroscopic bodies result from their structures as groups of atoms, but they are still real in a relational way. For example, cyanide is not intrinsically deadly, but it is still a real property when ingested by human beings.

POST-EPICUREAN ATOMISM

According to ancient historian Richard Carrier (he's not ancient, his history is), significant scientific progress was made in the period between Epicurus and Constantine, when Christianity took control of the Roman Empire. In personal correspondence, Carrier told me, "By influencing even non-atomist scientists and driving many of the debates (between atomist and non-atomist scientists), atomism was a major contributor to all the scientific progress in antiquity after Aristotle."[16]

Let me just briefly mention some of the scientific accomplishments of Greek and Roman scientists that occurred during this period. They are not recognized as well as they should be because the Church suppressed their writings due to their real or implied atheism.

Strato of Lampsacus (ca. 335–ca. 268 BCE) was the third director of the Lyceum founded by Aristotle. He wrote mostly on physics and disagreed with many of Aristotle's views, especially his teleology (final cause). He was an atomist, a materialist, and an atheist.

Ctesibius of Alexandria (fl. 285–222 BCE) founded the science of pneumatics, and his improved water clock was more accurate than any clock ever built until the pendulum clock invented by Christiaan Huygens in the seventeenth century.

Eratosthenes of Cyrene (ca. 276–ca. 195 BCE) invented the science of geography and was the first person to accurately estimate the circumference of Earth.

Archimedes of Syracuse (ca. 287–ca. 212 BCE) was one of the greatest scientists in the ancient world. His numerous achievements are well enough known that they need not be cataloged here.

Aristarchus of Samos (310–ca. 230 BCE) was the first known person to propose the heliocentric model of the solar system.

Hipparchus of Alexandria (ca. 190–ca. 120 BCE) developed trigonometry, discovered the precession of the equinoxes, and compiled the first comprehensive star catalog.

Hero (or Heron) of Alexandria (ca. 10–70 CE) was a geometer and inventor. He collected geometric formulas from a variety of ancient sources on the areas and volumes of various solid and planar figures. He also invented over a hundred pneumatic devices, that is, machines that work by air, steam, or water pressure. These included a fire engine, a wind organ, and a steam-powered engine that was the first known device to transform steam into rotary motion. He also proposed mechanical devices that used levers, wedges, pulleys, and screws.

Claudius Ptolemy (ca. 90–ca. 168 CE) developed the detailed

geocentric model of the solar system that enabled astronomers to predict the positions of planets, the rising and setting of stars, and eclipses of the sun and Earth's moon.

LUCRETIUS

Titus Lucretius Carus was a Roman citizen who lived around 100–50 BCE. While little is known of him, he wrote a 7,400-line poem, *De rerum natura* (*The Nature of Things*) that is considered one of the great works of the ages. It is also the most unusual of poems. It is not an epic tale of human or superhuman adventure. Nor is it myth or history, but a philosophical and scientific treatise written in Latin hexameter. Furthermore, the message is atheistic and materialistic, denying the existence of anything magical or supernatural, including an immortal soul, and proclaiming the evils of religion.[17]

The poem introduces few philosophical or scientific ideas original to the author but rather presents the worldview of Epicurus and the atomic theory as the master elaborated on what had been taught by Democritus. By putting the teachings of Epicurus in poetic form, Lucretius made them more palatable—indeed, he made them majestic and inspiring.

Many translations of *De rerum natura* now exist, although as we will see in the next chapter, that almost did not happen. In the seventeenth century, the great English poet John Dryden (1631–1700) translated selected portions, 615 lines out of 7,400, essentially rewriting them so they would be pleasing to the reader in English. He focused on subjects that appealed to him, such as the progress of love, the advantages of reason and moderation, and the inevitability of death. He ignored the philosophical passages, which offer the translator less freedom.[18]

I particularly enjoy these passages:

So when our mortal frame shall be disjoin'd
The lifeless lump uncoupled from the mind,
From sense of grief and pain we shall be free;
We shall not feel, because we will not be.

Nay, though our atoms should revolve by chance,
And matter leap into the former dance;
Though time our life and motion could restore
And make our bodies what they were before.

What gain to us would all this bustle bring?
The new-made man would be another thing;
When once an interrupting pause is made,
That individual Being is decayed.

We who are dead and gone, shall bear no part
In all the pleasures, nor shall feel the smart,
Which to that other Mortal shall accrue,
Whom of our matter time shall mold anew.

These passages recognize a fact that many scholars today fail to comprehend. Every event that happens in the universe is fundamentally reversible. The air in a room can rush out an opened door, leaving behind a perfect vacuum. All that has to occur is that the air molecules in the room, which are in random motion, be simultaneously moving in the direction of the door when the door is opened. That we never observe this to occur is purely a matter of chance. It is enormously unlikely, given the large number of molecules in the room, but technically not impossible.

Similarly, after a person dies, it is not impossible that her molecules reverse their directions and reassemble so she lives again. Here Dryden uses verse to explain more fully what he believed Lucretius intended in 3.850–51 when he said, as translated in prose, "It would not concern us at all, when once our former selves was destroyed."[19]

However, if our "selves" are solely composed of all the atoms in our bodies, most particularly our brains, then why wouldn't the full self be restored? Dryden seems to hold a dualistic view, certainly common in his time as now, when he talks about "the lifeless lump uncoupled from the mind." I don't think Lucretius meant that.

In what follows, I will provide some selected quotations from *De rerum natura* using a recent line-by-line translation in rhyme by A. E. Stallings that attempts to give some of the feel of the poetry of the work without departing too far from what Lucretius actually wrote.[20]

Let me begin very early in the poem, when Lucretius talks about how religion "breeds wickedness" and "gives rise to wrongful deeds." He uses the tale in the *Iliad* where Agamemnon sacrifices his daughter Iphigenia as his ships sail to Troy:

> With solemn ceremony, to the accompanying strain
> Of loud-sung bridal hymns, but as a maiden, pure of stain,
> To be impurely slaughtered, at the age when she should wed,
> Sorrowful sacrifice slain at her father's hand instead.
> All this for fair and favorable winds to sail the fleet along! —
> So potent was Religion in persuading to do wrong. (1.96–101)

The poem tells us to observe nature in order to eliminate religious darkness:

> This dread, these shadows of the mind, must be swept away
> Not by rays of the sun nor by the brilliant beams of day,
> But by observing Nature and her laws. And this will lay
> The warp out for us—her first principle: *that nothing's brought
> Forth by any supernatural power out of naught.* (1.146–50).

At this point, I want to correct a common misunderstanding. Ancient atomism has been interpreted, even to the present day, to imply that the existence of atoms was inferred by reason alone, thus providing a counterexample to the common view held by most scientists and philosophers of science that we can learn about the world only through observation. For example, in a 2012 op-ed piece in the *New York Times* titled, "Physicists, Stop the Churlishness," essayist Jim Holt criticizes the public disdain that several top contemporary physicists hold for philosophy.[21]

Holt quotes Richard Feynman as mocking "cocktail-party philosophers" for thinking they can discover things about the world "by brainwork rather than experiment." I have not been able to find the precise quotation, which Holt does not cite. However, Feynman does mention "cocktail-party philosophers" several times in chapter 16, volume 1, of his classic "Lectures on Physics."[22] Note that he did not say *professional* philosophers. In any case, Holt remarks, "Leucippus and Democritus . . . didn't come up with [the idea of atoms] by doing experiments."

I agree that physicists should not disparage philosophy, which performs a valuable service in clarifying and interpreting scientific results. However, I do not know of any professional philosophers today who claim that we can discover things about the universe by thought alone. Yet, Holt seems to think just that. If Holt is implying that knowledge of the universe can be obtained by thought and reason alone, he is surely at odds with the thinking of most scientists and philosophers.

Certainly Leucippus and Democritus, or any ancients, did not do the type of carefully controlled experiments that mark science today. Indeed, virtually all ancient philosophers, including Plato and Aristotle as well as the atomists, expressed distrust in the senses and believed that they could be overruled by reason. It was not until almost two millennia later that Galileo (1564–1642) reversed the inequality and established the rule of observation over pure thought (as well as revelation), which then became the governing principle of the scientific revolution that followed.

Still, even after that, philosopher Immanuel Kant (1724–1804), in his *Critique of Pure Reason*, argued that the mind had access to truths about the universe that did not depend on observation, what he called *synthetic a priori* knowledge. One of his major examples was our intuition that space is described by Euclidean geometry. We now know that other geometries exist and that Einstein used non-Euclidean geometry to describe space in his general theory of relativity.

While general relativity was certainly a remarkable achievement of the human intellect, it was pursued because of the failure of Newton's theory of gravity to explain certain observations, such as the precession of the perihelion of Mercury, and was only accepted after it had explained these and successfully predicted other observations.

No one knows how the atomists arrived at the idea of atoms, but they weren't just brains in a vat operating by thought alone. They were living, experiencing human beings.

The following quotation from Lucretius shows that at least he, writing almost three centuries after Democritus, sought empirical justification for the atomic model.

> Just in case you start to think this theory [atoms] is a lie,
> Because these atoms can't be made out by the naked eye,
> You yourself have to admit there are particles
> Which *are* but which cannot be seen . . . (1.165–69)

For example,

> Thus clearly there are particles of wind you cannot spy
> That sweep the ocean and the land and clouds up in the sky.
>   (1.277, 278)

Further, in book 2 he adds,

> There's a model, you should realize,
> A paradigm of this that's dancing right before your eyes—
> For look well when you let the sun peep in a shuttered room
> Pouring forth the brilliance of its beams into the gloom,

And you'll see myriads of motes all moving many ways
Throughout the void and intermingling in the golden rays. (2.112–17)
. . .
Such turmoil means that there are secret motions, out of sight,
That lie concealed in matter. For you'll see the motes careen
Off course, and then bound back again, by means of blows un-
    seen. (2.126–28)

This remarkable passage is suggestive of the motion that we now know as Brownian motion. Einstein and Jean Baptiste Perrin used Brownian motion in the early twentieth century to demonstrate conclusively the existence of atoms. French philosopher Gaston Bachelard (1884–1962) was of the opinion that observations involving dust provided the essential notion of atoms so that it was not just a product of pure thought:

Without this special experience, atomism would never have evolved into anything more than a clever doctrine, entirely specu-lative, in which the initial gamble of thought would have been justified by no observation. Instead, by virtue of the existence of dust, atomism was able to receive from the time of its inception an intuitive basis that is both permanent and richly evocative.[23]

However, in this case, the dust motes are also moved around by air currents. In the Brownian motion observed in other media, currents are negligible.

Continuing with the poem, Lucretius follows Epicurus in con-tradicting Aristotle on the existence of the void:

For if there were no emptiness, nothing could move; since it's
The property of matter to obstruct and resist,
And matter would be everywhere at all times. So I say
Nothing could move forward because nothing would give way.
    (1.331–36)

Here he also gives basically the definition of matter that I always state as follows: *matter is what kicks back when you kick it.*

Lucretius also anticipates modern physics in his view of time:

> As slavery, penury and riches, freedom, war and peace,
> Whatever comes and goes while natures stay unchanging, these
> We rightly tend to term as 'consequences' or 'events'.
> Nor does Time exist in its own right. But there's a sense
> Derived from things themselves as to what's happened in the past,
> And what is here and now, and what will come about at last.
> No one perceives Time in and of itself, you must attest,
> And something apart from things at motion and from things at
>   rest. (1.455–63)

The notion of time expressed above clashes with the common-sense view but is very suggestive of the meaning of time implied by the theories of relativity.

We saw above that Epicurus had introduced the idea that atoms randomly "swerve" to the side during their natural downward motion, so that they could interact with other atoms. Here's how Lucretius describes it:

> . . . when bodies fall through empty space
> Straight down, under their own weight, at a random time and
>   place,
> They swerve a little. Just enough of a swerve for you to call
> It a change of course. Unless inclined to swerve, all things would
>   fall
> Right through the deep abyss like drops of rain. There would be
>   no
> Collisions, and no atom would meet atom with a blow. (2.216–23)

Richard Carrier has listed twenty-two "predictions" made by Lucretius that have been verified by modern science. Carrier has also provided the exact locations with the poem.[24] Of course, these were not the type of precise, falsifiable predictions we see in science today, and they are, in some cases, stretching it a bit. Nevertheless, they illustrate that the successful prediction of atoms was not simply a lucky random occurrence that might have been made by an astrologer or an alchemist, but that it is part of the larger worldview that follows naturally from the assumption that everything is particles and the void. Now that the atomic model has been fully verified by modern science, that worldview is ready to be taken seriously.

Finally, I mentioned in the preface that the atomists anticipated evolution by natural selection. Lucretius talks about how, in the beginning, there were many freaks with various deformities that made them unable to reproduce or forage for food and so their species died off. You will get objections from some scholars that this was not really evolution, so I will just provide the following excerpt:

> Many kinds of creatures must have vanished with no trace
> Because they could not reproduce or hammer out their race.
> For any beast you look upon that drinks life-giving air,
> Has either wits, or bravery, or fleetness of foot to spare,
> Ensuring its survival from its genesis to now. (5.855–59)

THE ANTIATOMISTS

Aristotle

Aristotle had no sympathy for atomism. He refined the model attributed to the pre-Socratic Empedocles (ca. 490–430 BCE) that fire, earth, air, and water constitute the basic elements out of which everything else is formed.

Aristotle disagreed with the atomist view that observable qualities such as color and smell were "conventions" and that only atoms are real. Rather, he insisted these were intrinsic properties of bodies that had nothing to do with the observer. Aristotle's primary reason for rejecting atomism was his conviction that a void was logically impossible. Putting it in modern terms, he also thought that the natural speed of a body was inversely proportional to some resisting factor. Since the void has no resisting power, atomic speeds are infinite, which Aristotle considered absurd. It followed, then, that there is no void.

Aristotle's argument was based on his theory of motion, which was grossly wrong, where by "grossly wrong" I mean totally inconsistent with observations.

What were Aristotle's gross errors?

1. He failed to grasp the principle of inertia, in which a body in motion can remain in motion. He assumed that in order to move, a body must be pushed along by some agent or motive power.
2. He assumed there were two types of motion: "natural" and "forced." The basic elements fire and air move naturally upward, while earth and water move naturally downward. Celestial bodies move naturally in circles. The agent for these motions was Aristotle's "final cause," the teleological principle in which everything has a purpose toward which it naturally progresses. All other motion is forced.

These views led Aristotle to believe that atoms and the void is necessarily a false theory. He reasoned that if an object were placed all by itself in a void, there could be no natural motion since there was no up or down. The object would not know where to go. Furthermore, being alone, it had no forces acting on it. It followed that motion in a void is impossible.

Aristotle's second misconception led to a third error that would also have far-ranging consequences on scholarship in the Middle Ages. Aristotle concluded that a different set of dynamical principles governed Earth and the heavens. Here he actually retrogressed from the views of the Milesian philosophers. It was only with the discoveries of Kepler, Galileo, and Newton two millennia later that it was established that physics is universal.[25] It is surely significant that the scientific revolution of the seventeenth and eighteenth centuries occurred outside the Church-controlled universities in Europe where Aristotelian scholasticism was being taught as dogma.

The Stoics

Pullman identifies the physics doctrines of the Stoic philosophers as "the clearest expression of opposition to the teachings of the atomists."[26] Stoicism was founded in Athens by Zeno of Cittium in the third century BCE and had adherents well into the Common Era, including the Roman emperor Marcus Aurelius (121–180). Stoicism fizzled out in 529 when the emperor Justinian I (482–565) shut down all philosophical schools so that Christianity would have no opposition.

As we have already seen in the case of atomism, a philosophy of the nature of the universe affects thinking in other areas such as religion and morality. Such was the case with the Stoics. While the atomists divided the universe into discrete parts separated by empty space, the Stoics viewed it as a continuum without any void. While the atomists believed in an impersonal universe, the Stoics were pantheists, holding that the universe was an active, reasoning substance.

In his work *De Natura Deorem* (*The Nature of the Gods*), the great Roman statesman and philosopher Cicero (106–43 BCE) explains the Stoic view this way:

> The universe itself is god and the universal outpouring of its soul;
> it is this same world's guiding principle, operating in mind and
> reason, together with the common nature of things and the totality
> which embraces all existence.

Unlike the atomists, who believed that material atoms are unlimited and eternal, the Stoics' entire world was finite and corruptible. Unlike the atomists, who believed that chance played a role in the evolution of the world, to the Stoics, everything was predetermined by the ultimate organizing force. Out of this belief came the well-known characteristic that today is commonly called *stoicism,* the acceptance of fate. Atomism fully accepts human free will, although to the atomists it is the result of the random swerve, which is not exactly what Christianity and other religions think of as free will.[27]

Like so many people today, the Stoics could not see how the complexity of the world could arise by chance. Cicero scoffed at the notion, comparing it with tossing a vast quantity of the twenty-one letters of the Latin alphabet on the ground and having it produce Ennius's *Annals.*[28]

Cicero also did not accept the Epicurean notion of the swerve, as described previously. What causes an atom to deviate from its path? Once again, we see how difficult it is to accept the notion that not everything requires a cause, that some things simply happen by chance.

The Neoplatonists

The Neoplatonists constituted the final group of early antiatomists. Their school was founded by Plotinus of Lykopolis (205–270 CE). His collected writings appear in the six *Enneads* and were a great influence on Christian theology as incorporated by Augustine of Hippo in the fifth century CE.

Although the Neoplatonists did not support stoicism, they agreed that it was absurd to think the world could be the result of spontaneity and chance, that everything simply arose from the movements of atoms. In *Enneads* 2.1, Plotinus asks,

> What motion of atoms can one attribute to the actions and passions of the soul? . . . What movements of atoms stir the thought of the

geometer, the arithmetician, or the astronomer? What movements are the source of wisdom?

Plotinus anticipates many modern theological arguments when he insists the following in *Enneads* 4.4:

> It is impossible for the association of material bodies to produce life and for things devoid of intelligence to engender intelligence. . . . For there would be no composite bodies and not even simple bodies in the reality of the world, were it not for the pervasive soul of the universe.

As we will see, in every important respect, the atomists were so right and the antiatomists so wrong.

2

ATOMS LOST
AND FOUND

I am an Epicurean.

—Thomas Jefferson

ATOMISM IN EARLY CHRISTIANITY

Atomism isn't mentioned in the Bible, although a reference to Epicureans during a visit by Paul to Athens can be found in the Acts of the Apostles. "What will this babbler say," they asked (Acts 17:18). (Aside: When I grabbed my King James Bible to check this reference, it opened right to the page.) However, there can be little doubt that atomic philosophy, if not atomic physics, conflicts with Christian teaching—not only with scriptures but also with the teachings of Plato and Aristotle that had great influence on Church theologians.

In the fifth century, Augustine of Hippo was quite explicit in rejecting any notion of the primacy of matter:

> Let those philosophers disappear, who attributed natural corpore-
> al principles to the intelligence attached to matter, such as Thales,
> who refers everything to water, Anaximenes to air, the Stoics to
> fire, Epicurus to atoms, that is to say, to infinitely small objects that
> can neither be divided nor perceived. Let us reserve the same fate
> to all the other philosophers, who are too numerous to name and
> who claimed to have found in simple and combined substances,

lifeless or living, indeed in what we call material bodies, the cause and principle of things.[1]

Bernard Pullman provides a long list of disagreements between atomism and Christianity.[2] I will mention just a few of the most important.

To avoid confusion, here and in the remainder of this book I will refer to the atomists' notion of an eternal, uncreated universe containing multiple worlds with the modern designation *multiverse*. The term *universe* will then generally refer to our universe and other individual islands within the multiverse. Clearly, an eternal multiverse is at odds with the Christian belief in creation. Furthermore, either our universe, our galaxy, our sun, our Earth, and our form of life are unique manifestations, or Jesus had to die on the cross countless times in all those universes, on each and every inhabited planet. Neither case makes much theological sense. We humans like to think we are unique, and theology supports that view, but then why would God create so many other universes, other galaxies, other stars, other planets, and (most likely) other forms of life?

Christian theology also has a problem with the nature of time. If God were perfect, unchanging, and eternal, why would he have made a change in creating the world? Augustine, following Plato, thought he had solved the problem by saying God created time along with the universe. The atomists did not have to trouble themselves over the question because they did not regard time as an element of reality but simply as a relation between events. This is also a modern concept. As Einstein said, "Time is what you read on a clock."[3]

Another area of disagreement concerned the atomists' attributing to chance the formation and evolution of worlds. To Christians, divine providence alone determines the fate of the universe. Recall from chapter 1 that this was also an area of disagreement with the Stoics, who regarded everything as predetermined by fate.

Obviously, the atomists' view of the soul as material and mortal is unacceptable to Christians.

Throughout history, Christian theologians attacked the moral teachings of Epicurus, often misrepresenting them, as did the Stoics. Augustine wrote that "the pleasure advocated by Epicurus is the realm of beasts only. . . . [He] summons from his gardens a throng of his inebriated disciples to his rescue, but only to search frantically what they can tear apart with their dirty fingernails and rotten teeth."[4]

ATOMISM IN THE MIDDLE AGES

While the Church did its best to suppress the writings of the Epicureans, medieval scholars of Christianity, Judaism, and Islam showed sufficient interest that knowledge of the philosophy and physics of atomism survived in their writings.[5] Not all these chroniclers were necessarily supporters or proponents of atomism. None accepted its atheistic elements, but some found its physics congenial. The very fact that atomism conflicted with both Aristotle and the dominant monotheisms of the age made it a fit subject for philosophical and theological commentary, even if the goal was to refute that heresy.

While scriptures and the Church provided authority, reason was recognized as an auxiliary means by which God's laws could be accessed. Thus, with this caveat, some twelfth-century thinkers such as Adelard of Bath (1075–1150), Thierry of Chartres (ca. 1100–ca. 1150), and William of Conches (ca. 1090–1154) thought that the physics of atoms made sense. However, they were still a small minority.

Support of atomism was often associated with opposition to Aristotle. In the fourteenth century, William of Ockham (ca. 1288–1348), of "Ockham's razor" fame, was highly critical of Aristotle and claimed that matter could be reduced to "elementary particles." He also agreed with the atomists that the universe was infinite and eternal. The Church condemned these theses in 1340.

Nicholas of Autrecourt (ca. 1299–1369) also defended atomism and repudiated Aristotle. However, he did not accept the Democritus-Epicurus view of the soul and considered it to be composed of two immortal spirits he called *intellect* and *sense*.

The only atomists within medieval Judaism were members of a schismatic sect called the Karaites. They were condemned by the most influential Jewish thinker of the time, Moses Maimonides (1135–1204), who opposed atomism as well as other Karaite doctrines. In his most famous work, *The Guide of the Perplexed*, Maimonides mentions Epicurus and, in one long sentence, tells us why we should ignore him:

> As for those who do not recognize the existence of God, but who believed that things are born and perish through aggregation and separation, according to chance, and that there is no being that rules and organizes the universe—I refer here to Epicurus, his sect and the likes of him, as told by Alexander—it serves no purpose for us to speak about those sects; since God's existence has been established, and it would be useless to mention the opinions of individuals whose consciousness constructed their system on a basis that has already been overthrown by proofs.[6]

While Christendom was mired in the Dark Ages, Islam was going through its golden age. Scholarship flourished throughout the vast empire that had been conquered by the followers of Muhammad.[7] Maimonides traveled extensively throughout these lands and wrote about what he learned.

Within Islamic scholarship, there existed a discipline called the *Kalām* that practiced the kind of theological rationalism I mentioned earlier, where reason is used to develop knowledge of God. In *The Guide of the Perplexed*, Maimonides described in some detail Arabic atomism as expressed in the Kalām.[8]

Basically, Arabic atomism followed Greek atomism in asserting that the universe is composed of miniscule indivisible particles that combine to give material substances. They also affirmed the existence of void. They viewed time as discontinuous, composed of instants.

As with Christian atomism, grave differences with the atomism of Democritus and Epicurus unsurprisingly arose when it came to God and the soul. In Islamic atomism, while everything is composed of atoms, these atoms are not eternal. In fact, they exist for only an instant and are continually re-created by God. Nothing depends on what went on before. There are no natural causes; God is the only cause. God has complete freedom and is responsible for every event that happens in the universe down to the finest detail.

Maimonides did not think much of the idea of perpetual creation. For example, he pointed out that long after God causes a person to die, he has to keep re-creating the leftover atoms such as those in teeth that survive for thousands of years.

Not all Arabic scholars went along with Kalām, notably the scholar best known in Europe, Averroes (Ibn Rushd, 1126–1198). However, as Pullman notes, "Among the three great monotheistic religions in the West, Islam [Kalām] was the first to proclaim that faith in a unique God, master of the universe, is entirely compatible with a corpuscular conception of the structure of matter. . . . That one can accept an atomic vision of the world regardless of one's position vis-à-vis God."[9]

POGGIO AND LUCRETIUS

Meanwhile, atomism continued to be strongly opposed in Christendom. Although suppressed by the Church, a copy of *De rerum natura* luckily survived intact and, after being rediscovered in the fifteenth century, played no small part in the Renaissance and the scientific revolution that followed on its heels.[10]

Literary scholar Stephen Greenblatt has told the fascinating story of the rediscovery of *De rerum natura* in *The Swerve: How the World Became Modern.*[11] The central figure is Gian Francesco Poggio Bracciolini (1380–1459), who had served as apostolic secretary to five popes. Poggio had a deep knowledge of Latin, had beautiful handwriting, and enjoyed a long career as an influential layman within the Catholic

Church bureaucracy. In 1417, he was out of a job after having served Baldassare Cossa, Pope John XXIII (ca. 1370–1419), the "antipope" who was deposed, stripped of his title, name, and power in 1413.[12]

Poggio was one of a group of scholars of the period called *humanists* who pored over classical Roman texts and sought out missing manuscripts. Relieved of his duties with the fall of Cossa, Poggio had time to follow this passion. At the time, monks were the major book preservers, and so monasteries were the first place to look for the desired texts. Poggio was drawn to a monastery in central Germany, the Benedictine Abbey of Fulda, founded in 744, which he had heard held a cache of old manuscripts. There he found a treasure of works by ancient Roman authors unknown to him and his fellow humanists. According to Greenblatt, even Poggio's smallest finds were highly significant.

However, these were eclipsed by his discovery of *De rerum natura*, a work more ancient than any of the others he uncovered in Fulda.[13] Poggio was well aware of Lucretius, who had been mentioned by Cicero and Ovid. Ovid wrote: "The verses of sublime Lucretius are destined to perish only when a single day will consign the world to destruction."[14] According to Greenblatt, *De rerum natura* also influenced Virgil. Greenblatt says: "Virgil's great epic, the *Aeneid*, was a sustained attempt to construct an alternative to *On the Nature of Things*: pious, where Lucretius was skeptical; militantly patriotic, where Lucretius counseled pacifism; soberly renunciatory, where Lucretius embraced the pursuit of pleasure."[15]

The monks in Fulda would not part with the manuscript, so Poggio arranged for a scribe to make a copy. After receiving the copy in Constance, he sent it to his friend and fellow humanist Niccolò Niccoli (1364–1437) in Florence, Italy, who then made another copy in his own elegant cursive script, which ultimately developed into italic type. That copy, Codex Laurentianus 35.30, resides today in Florence in the beautiful Laurentian Library designed by Michelangelo for the Medici family. I was able to personally view the manuscript on March 27, 2012.[16] Figure 2.1 shows the first page.

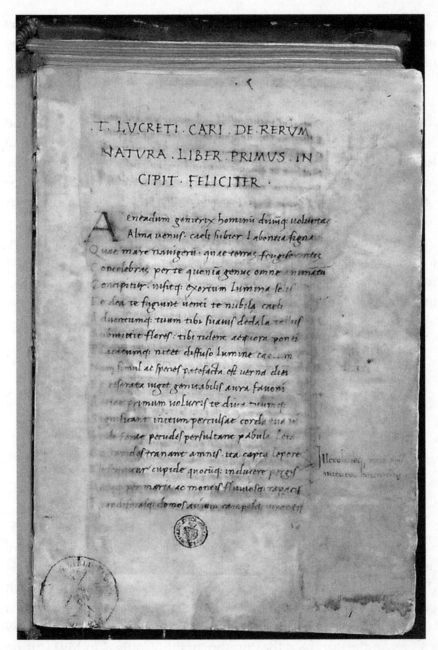

Figure 2.1. The first page of Lucretius's *De rerum natura*, copy by Niccolò Niccoli. (Photograph courtesy of the Laurentian Library, Florence, Italy.)

Many more copies were spawned from these two, including one by the notorious Niccolò Machiavelli (1469–1527). That copy resides in the Vatican Library, MS Rossi 884.

Poggio and his contemporaries were not overly alarmed by the atheism in *De rerum natura* or, more precisely, by the indifference of the gods. After all, Lucretius died a half century before Christ and so had no opportunity to learn the "truth." They approved of what Lucretius saw as the absurdity of the pagan practices of his contemporaries. Greenblatt points out, "even many modern translations of Lucretius' poem into English reassuringly have it denounce as 'superstition' what the Latin text calls simply *religio*."[17]

Still, the ideas uncovered in the poem had lain undiscovered for a thousand years and were bound to upset conventional thinking for no other reason than their strong unorthodoxy. Greenblatt eloquently summarizes the Epicurean message:

> It is possible for human beings to live happy lives, but not because they think that they are the center of the universe or because they fear the gods or because they nobly sacrifice themselves for values that purport to transcend their mortal existence. Unappeasable desire and the fear of death are the principle obstacles to human happiness, but the obstacles can be surmounted through the exercise of reason. . . . All speculation—all science, all morality, all attempts to fashion a life worth living—must start and end with a comprehension of the invisible seeds of things: atoms and the void and nothing else.[18]

At the time Machiavelli was making his copy of *De rerum natura* in Florence, the notorious Dominican friar Girolamo Savonarola (1452–1498) was ruling the city with religious fanaticism, which eventually led to his excommunication and execution. Savonarola had spoken out against atomism, so Machiavelli wisely kept his copy a secret, and it survived the "Bonfire of the Vanities" in 1497 during which the followers of Savonarola burned books and other "objects of sin." It was not until 1961 that Machiavelli's handwriting was identified and the copy in the Vatican was conclusively attributed to him.

Poggio was also able to avoid any taint of atheism resulting from his role as the discoverer of *De rerum natura* by separating the poetry from the message. Still the message circulated relatively freely until 1556 when the Florentine synod prohibited the teaching of Lucretius in schools. Nevertheless, this did not halt the printing of the poem in Italy and elsewhere. By then, editions had appeared in Bologna, Paris, and Venice, and a major edition in Florence had attracted much attention. Attempts to place *De rerum natura* on the Catholic Church's *Index Librorum Prohibitorum* (*Index of Prohibited Books*) failed, and Catholic intellectuals were allowed to discuss Lucretius as long as they treated it as a pagan fable.

Both Erasmus (ca. 1466–1536) and Thomas More (1478–1535) attempted to reconcile Epicurus and Lucretius with Christian thinking. More was, of course, the English statesman and scholar, the celebrated "Man for All Seasons,"[19] who was beheaded by King Henry VIII for refusing to take an oath acknowledging the supremacy of the Crown in the Church of England. He was sainted in 1935.

Although admired for his steadfast loyalty to the Roman Catholic Church, More was a religious fanatic who wore a hair shirt and whipped himself until blood flowed. While he was Lord Chancellor of England from October 1529 to July 1535, More saw to it that six heretics were burned at the stake (not an unusual number at the time).

Still, More described himself as a "Christian humanist," and his best-known work, *Utopia,* is a novel in Latin about an imaginary island where people live in peace under orderly social arrangements, free of the misery and conflict that then existed in Europe. More's Utopians were inclined to believe "that no kind of pleasure is forbidden, provided no harm comes of it." However, while the citizens of Utopia are encouraged to pursue pleasure, those who think that the soul dies with the body or who believe that chance rules the universe were to be arrested and enslaved.[20]

So, while More adopted Epicurus's "pleasure principle," the seeking of pleasure had to be done under strict limitations. While people could worship any god they pleased, they could not follow Epicurus and Lucretius and worship no god or doubt the immortality of the soul.

While we are talking about sixteenth-century figures who were executed for their beliefs, let us not forget Giordano Bruno. He had a tangled philosophy that included Epicureanism. According to Greenblatt, Bruno "found it thrilling that the world has no limits in either space or time, that the grandest things are made of the smallest, that atoms, the building blocks of all that exists, link the one and the infinite."[21]

Bruno championed Copernicanism at a time when the notion that Earth moves around the sun was unpalatable to both the Church and academic scholarship wedded to Aristotle. Bruno even went further than Copernicus in saying that the sun was not the center of the universe either, but that there was no center. Here again we find a brilliant centuries-old intuition, also held by the atomists, that would not be confirmed until the twentieth century, in this case when Einstein's 1916 general theory of relativity was applied to cosmology.

Bruno was burned at the stake in Rome on February 17, 1600.

GASSENDI

Pierre Gassendi (1592–1655) was a transitional figure who played an important role in moving the intellectual world from medieval thinking into the scientific age and helped make atomism a crucial element in that revolution. Gassendi was a French philosopher, priest, scientist, and classical scholar—a contemporary of René Descartes (1596–1650), Galileo, and Kepler.[22]

Although as a priest he adhered to the theological elements of Church doctrine, Gassendi was a strict empiricist who insisted that knowledge of the external world is built solely on sensory evidence. He did not just talk about observations and experiments, he performed them.

Using telescope lenses provided by Galileo, to whom he wrote letters of support, Gassendi made numerous observations that helped establish the validity of Kepler's laws of planetary motion. In 1631, he became the first to observe a planetary transit of the sun (in this case, Mercury), providing strong confirmation of the Copernican model. This made possible the first estimates of distances between Earth, the sun, and the planets. Nevertheless, he was careful to say that other models were possible pending further data. Gassendi's other observations included sunspots, eclipses of the sun and moon, and the handles of Saturn that would later be identified as rings. He denounced astrology since it had no empirical support.

In physics, Gassendi studied free fall, measured the speed of sound, and showed that atmospheric pressure was lower on a mountaintop than at sea level, thus adding to the evidence that a void was possible. He performed a very significant experiment in which a stone is dropped from a moving ship, showing that the stone maintains the horizontal speed of the ship. Galileo had suggested this as a thought experiment to illustrate his principle that motion is relative. The best way to see this is from the point of view of an observer on the ship. That observer will see the stone drop straight down, whether the ship is moving or at rest. This was crucial in answering the quite-legitimate question addressed to Galileo: If Earth moves, why don't we notice it? The answer: Because there is no difference between being at rest and being in motion at constant velocity.[23]

Gassendi also made a major step toward the law of inertia as a result of this experiment. Galileo had thought of inertial motion as fundamentally circular. Gassendi realized that the natural motion of a body is in a straight line.

On the philosophical side, Gassendi was part of the growing movement that chipped away at Aristotelian scholasticism, which had dominated the universities of Europe for centuries. Recall that Aristotle claimed a duality of matter and immutable "essences." Descartes made the same distinction with matter and "mind" to which most people still cling today. Gassendi maintained that regardless of whether there are any essences, we have no way of knowing about them. He also wrote a criticism of Descartes's *Meditations* and the reasoning behind *cogito, ergo sum*, basing it on empirical arguments. This generated a sharp reply by Descartes and further public debate between the two. Descartes had noted that we can only perceive appearances. Gassendi agreed, adding that appearances are all we can know about, which rules out any knowledge of essences.[24]

Gassendi's greatest achievement, however, was in rehabilitating Epicurus and bringing the atomic model to center stage in the new physics. Here he had to resolve a clear conflict with his empiricism. How can we say atoms exist if we can't see them?

While Gassendi agreed that we can't know anything for certain, he said we can still use indirect empirical evidence to support hypotheses about the invisible. Atomism is presented as the most likely hypothesis, what we now call "inference to the best explanation." He claimed as such evidence the structures we see with a microscope, such as crystals.[25]

Gassendi translated Diogenes Laertius's book 10 on Epicurus from *The Lives of the Eminent Philosophers* into Latin, along with ample commentary, in *Animadversiones*, published in 1649. While he followed the ancient atomists on the basic reality of atoms and the void, the priest Gassendi still asserted that they were put there by God.

Gassendi made further advances to atomism by proposing that light and sound are particulate. He provided speculative atomic accounts of planetary motion, of chemical and biological phenomena, and even of psychology. While none of his specific proposals, except the atoms of light, have withstood the test of time, they served to establish the notion that everything might someday plausibly be explained solely by atoms and the void.

Gassendi influenced any number of seventeenth-century scholars, including Robert Boyle (1627–1691) and John Locke (1632–1704). One interesting story is how the theory of nerve transmission developed by Thomas Willis (1621–1675) and Isaac Newton was based on a proposal by Gassendi. This supplanted the Cartesian model that separated the mind from the nervous system and instead treated nerves as communication lines with the brain.[26]

Newton adopted several of Gassendi's ideas, such as the particulate nature of light that was in opposition to the wave theory proposed by Robert Hooke (1635–1703) and Christiaan Huygens (1629–1695). In short, Pierre Gassendi was an important, insufficiently recognized contributor to the scientific revolution that followed.

3

ATOMISM AND THE SCIENTIFIC REVOLUTION

It seems probable to me that God in the beginning formed matter in solid, massy, hard, impenetrable, moveable particles, of such sizes and figures, and with such other properties, and in such proportion to space, as most conducted to the end for which he formed them.

—Isaac Newton, Query 31, Book 3,
Opticks (1704)

THE NEW WORLD OF SCIENCE

Historians still debate the causes of the dramatic upheaval in human thinking that took place in the seventeenth century called the scientific revolution. Some, a minority, even dispute that it was a revolution. The common wisdom is that the scientific revolution replaced the magical thinking and superstition of the medieval age, in which knowledge was primarily based on revelation and sacred authority, with rational thought founded on observation and experiment. However, most historians today say, as experts on anything always say, the truth is more complex.

The natural philosophy that originated in Greece continued into the Middle Ages, mainly in the Arabic empire.[1] Meanwhile, most secular intellectual endeavor in Europe sank into decline. Still, medieval Europe was not totally absent of scholars who recognized the importance of observing nature. In my previous book I described how these scholars, notably Augustine of Hippo, viewed science as the handmaiden of religion by providing knowledge of God's creation.[2] Several modern historians, notably the French physicist and devout Catholic Pierre Duhem (1861–1916), claimed to see continuity between medieval scholarship and what is generally referred to as the "paradigm-shift" of seventeenth-century science.[3]

Some apologists have even gone so far as to argue that Christianity was the source of modern science.[4] However, this hardly jibes with the historical fact that Greece and Rome were well on their way to science, as we know it today, until the fourth century, when the emperor Constantine (272–337) empowered Christianity and it became the state religion. The Catholic Church then proceeded to systematically eliminate alternative religions of every variety, including various polytheisms and any competing monotheisms. These were suppressed throughout the empire, along with any scintilla of freethinking.[5]

The Dark Ages roughly spanned the thousand-year period from 500 to 1500, when the Roman Catholic Church dominated the Western Empire. They ended only after the Renaissance and Reformation undermined the Church's authority. During the period of Church rule, science not only failed to advance but was also set back. Surely, this is no accidental coincidence; although there is no doubt the Dark Ages were a product of many other forces and not just Church dogmatism.[6]

Nevertheless, recent scholarship has confirmed that a few scholars in the fourteenth century had already developed several of the basic mathematical principles of motion that were later to be rediscovered and fully implemented by Galileo Galilei (1564–1642) and Isaac Newton (1642–1727). A French priest, Jean Buridan (ca. 1300–1358), introduced the concept of *impetus* by which a projectile remains in motion unless acted on by a contrary force. He defined impetus as the quantity of matter in a body multiplied by its velocity. Today we call this *momentum* (mass × velocity). However, Buridan regarded impetus as the *cause* of motion, while the mechanics of Galileo and Newton recognized it as a *measure* of motion that requires no cause.[7]

Also in the fourteenth century, a group of English scholars at Merton College, Oxford, called the "Oxford Calculators," were developing the mathematics of uniformly accelerated bodies, including the law of falling bodies, that is usually attributed to Galileo.[8] The Oxford Calculators made a distinction between dynamics, which is the cause of motion (or, as we now say, changes in motion), and kinematics, which is the effect. They also made a clear distinction between velocity and acceleration. The group included Thomas Bradwardine (ca. 1290–1349), who later became archbishop of Canterbury. Like Buridan, the Oxford Calculators were mostly churchmen.[9]

Some of the other factors that are normally attributed to the rise of science but were already present in medieval scholarship include: (1) the application of mathematics to physics and (2) the use of the experimental method.[10]

However, these advances did not have significant impact until Galileo pointed his telescope to the heavens and performed experiments on the motion of bodies that demonstrated the superiority of observation over revelation or pure reason. These observations, of the sky and in the laboratory, revealed a physical world that bore little resemblance to the commonsense notions of the rest of humanity. As early twentieth-century philosopher Alexander Koyré put it:

> What the founders of modern science, among them Galileo, had to do, was not to criticize and to combat certain faulty theories, and to correct or replace them by better ones. They had to do some-

thing different. They had to destroy one world and to replace it by
another. They had to reshape the framework of our intellect itself,
to restate and to reform its concepts, to evolve a new approach to
Being, a new concept of knowledge, a new concept of science—
and even to replace a pretty natural approach, that of common
sense, by another that is not natural at all.[11]

Koyré asserted that the modern attempt by Duhem and others to
"minimize, or even to deny, the originality, or at least the revolutionary character, of Galileo's thinking" and to claim continuity
between medieval and modern physics "is an illusion."[12]

What Koyré refers to as "natural" above is not thinking mate-
rialistically, as we make the connection today, but rather thinking
commonsensically. Common sense is the human faculty for forming
concepts based on everyday experience, such as believing the world
is flat. The everyday experiences humans had until the seventeenth
century led them to view themselves at the center of the universe.
The experience of looking through a telescope resulted in a radical
new concept of the universe, one in which Earth moves around the
sun and the universe has no center. If anything, the history of science
is marked by the continual overthrow of common sense. Today,
over a century after they were first proposed, we still have trouble
reconciling relativity and quantum mechanics with common sense.

As we have seen, Aristarchus of Samos proposed the heliocentric model of the solar system two centuries before the Common
Era, but this knowledge was largely forgotten in the Dark Ages.
When reintroduced in the sixteenth century by Nicolaus Copernicus
(1473–1543) and advanced by Galileo, it was not immediately established as any better than the ancient geocentric model of Claudius
Ptolemy.

I need not repeat the oft-told story of Galileo's trial by the
Inquisition in 1615.[13] He had been ordered by Church authorities
not to teach the Copernican model as fact but simply as a calculational tool. Convicted and sentenced to (a very comfortable)

house arrest for life, and technically forbidden to do any further science, Galileo nevertheless proceeded to lay the foundation for Newtonian mechanics in his *Discourse on the Two New Sciences*, published in Holland in 1638.

GALILEAN RELATIVITY

Commonsense experience leads us to take for granted that we can tell when we are moving and when we are at rest. Galileo was questioned, quite reasonably, if Earth moves, why don't we notice it? His answer became one of the most important principles of the new physics: motion is relative.

A skeptical cardinal might have proposed the following experiment to Galileo: "Go up to the top of the Tower of Pisa and drop a rock to the ground. Using your own formula $h = gt^2 / 2$, where $g = 9.8$ meters per second per second (the acceleration due to gravity) and $h = 57$ meters (the height of the tower), the rock will take $t = 3.4$ seconds to reach the ground. [I have converted the units he would have used to the metric system]. If, as you claim, Earth is moving around the sun at 30 kilometers per second, then the rock should fall 102 meters away from the base of the tower because Earth will have moved that far in that time. The rock lands at the base, which proves Earth cannot be moving."

Galileo would have insisted that, based on his telescopic observations, Earth moves (*"Eppur si muove"*) and, based on his experiments, the rock drops at the base of the tower. Thus, our theory of motion must accommodate those facts.

Here was perhaps Galileo's greatest contribution to the scientific revolution, establishing once and for all the superiority of observation over theory, especially those theories based on authority. Indeed, it is our reasoning—and not our observations—that is to be mistrusted, contrary to the teachings of medieval theologians who viewed observation as unreliable and Church authority as final.

So here's how Galileo solved the problem of why we don't notice Earth's motion. He introduced what we now call the *principle of Galilean relativity*. Let me state this principle in an updated, "operational" way that allows us to see exactly what it means and how it directly applies to the cardinal's proposed experiment.

The Principle of Galilean Relativity

There is no observation you can perform inside a closed capsule that allows you to measure the velocity of that capsule.

In the cardinal's experiment, Earth is essentially a closed capsule, since we are performing the experiment at the Tower of Pisa and not looking outside that environment. Thus, we cannot detect the Earth's motion by this experiment.

The principle of Galilean relativity implies that there is no observable difference between being at rest and being in motion at constant velocity. Today we have an advantage, not available prior to the 1950s, of flying in jetliners where we can hardly distinguish between being in motion and being at rest, except during takeoff, landing, and in turbulent air. In those exceptions, what we experience is a change in velocity—acceleration—and not motion itself.

[Technical note: the terms *speed* and *velocity* are often used interchangeably in normal discourse. In physics, the *velocity* **v** is the time rate of change of position and is a *vector*. That is, it has both a magnitude and direction and requires three numbers to specify. We use boldface type to designate familiar three-dimensional vectors. The *speed v* is the magnitude (or length, when drawn to scale on paper) of the velocity vector, indicated conventionally in italic script.]

When we are flying in an airplane, we are said to be in the plane's *frame of reference*. Someone standing on the ground is in Earth's frame of reference. We will have much occasion as we move the discussion into modern physics to talk about frames of reference.

Observers in different frames of reference often see things differently. If you are on a boat moving at constant speed down a river and you drop an object from the top of the mast, you will see it fall straight down to the base of the mast. Someone standing on shore, in a different reference frame, will see the object fall along a parabolic path. However, note that the two of you will still witness the same result, namely, the object landing at the base of the mast. In the time it took the object to fall to the deck, the boat will have moved ahead a certain distance; so, to the observer on shore, the object has a horizontal component of velocity exactly equal to the velocity of the boat. To the observer on the boat, that horizontal component is zero.

We can see how this follows from the principle of Galilean relativity. Suppose that instead of standing on deck, you are below in the hold, which has no portholes. You are not aware if the boat is moving or not, so you decide to find out by dropping a coin from your hand to the deck of the hold. If the coin does not land at your feet but some distance away, you will have detected that the boat is moving by an observation made solely inside the hold. This would violate Galileo's principle. You would have detected your motion inside a closed capsule. Instead, the coin will drop to your feet exactly as it would if the boat were tied up at the dock, verifying the principle of Galilean relativity.

THE *PRINCIPIA*

The heliocentric model was eventually accepted based on Galileo's telescopic observations but also because it proved superior to the Ptolemaic model once the data improved, thanks to Tycho Brahe (1546–1601) and Johannes Kepler (1571–1630). Kepler inferred from his own careful observations and from those of Brahe that the planets move around the sun in ellipses rather than circles. He proposed three laws of planetary motion.

Kepler's Laws of Planetary Motion

1. The orbits of planets are ellipses with the sun at one focus.
2. A line from the sun to the planet sweeps out equal areas in equal times.
3. The square of the orbital period is directly proportional to the cube of the semi-major axis of the orbit.

In January 1684, physicist Robert Hooke (1635–1703) was sitting in a London coffeehouse along with architect Christopher Wren (1632–1723) and astronomer Edmund Halley (1656–1742). They started talking about gravity. Halley asked if the force that keeps the planets in orbit could decrease with the square of distance. His companions both laughed. Wren said it was easy to reach that conclusion, but it was quite another thing to prove it. The boastful Hooke said he had proved it years ago but never made it public. Wren challenged Hooke to produce the proof in two months, but Hooke never did.[14]

Impatient with Hooke's failure to provide his proof, that summer, Halley went to visit Isaac Newton in Cambridge, whom he barely knew at the time, and asked the Lucasian Professor what the curve of a planetary orbit would be if gravity were reciprocal to the square of its distance to the sun. Newton responded immediately that it would be an ellipse, as Kepler had observed. Halley asked Newton how he knew that, and Newton replied, "I have calculated it."[15] Newton rummaged around, but he could not find the proof among his papers and promised to work it out again.

Unlike Hooke, Newton kept his promise. Three months later he sent Halley a nine-page treatise presenting the proof. This so impressed Halley that he personally funded the publication on July 5, 1687, of the greatest scientific work of all time, *Philosophiae Naturalis Principia Mathematica* (*Mathematical Principles of Natural Philosophy*). (Maybe you think Darwin's *On the Origin of Species* was greater, but let's not argue about it.)

Principia presented Newton's laws of motion and his theory of universal gravitation, from which Newton derived Kepler's laws of planetary motion. But *Principia* did much more. It provided the framework for the remarkable scientific achievements that followed. Perhaps the most remarkable was Halley's comet. Halley had determined that the comets that appeared historically in 1456, 1531, 1607, and 1682 were the same body, and he used the new physics to predict that it would reappear in 1759. When it did, after Halley's and Newton's deaths, few could any longer dispute the enormous power of the new science.

Three hundred years after it was glimpsed by Jean Buridan, another fundamental principle was carved in stone by Newton.

The Principle of Inertia

A body in motion with constant velocity will remain in motion at constant velocity unless acted on by an external force.

Usually this is accompanied by a similar statement for a body at rest, but by the principle of relativity there is no difference between being at rest and being in motion at constant velocity: rest is just "motion" at zero velocity in some reference frame (you can always find such a frame). So the added statement is redundant.

The law of inertia is the first of Newton's three laws of motion. All three boil down to another, more general principle, that is, the principle of conservation of momentum.

The Principle of Conservation of Momentum

The total momentum of a system of particles will remain fixed unless acted on by an outside force

where force is defined as the time rate of change of momentum in Newton's second law. Newton's third law, "for every action there is an equal and opposite reaction," also follows from conservation of momentum.

As just noted, the momentum **p** of a particle is its mass m times its velocity **v**, that is, $\mathbf{p} = m\mathbf{v}$.[16] Momentum is a vector whose magnitude is mv, where v is the speed, and whose direction is the same direction as the velocity vector. The total momentum of a system of particles is the vector sum of the individual particle momenta. Particles in a system not acted on by an external force can collide with one another and exchange momenta, as long as the total momentum remains fixed.

PARTICLE MECHANICS

Newton correctly inferred that gravity was not important in the mutual interaction of corpuscles and that other forces, such as magnetism and electricity, came into play there. He rejected the primitive notions of hooks or any "occult" quality holding atoms together to form composite bodies. He wrote:

> I had rather infer from their cohesion, that their particles attract one another by some force, which in immediate contact is extremely strong, at small distances performs the chemical operations above-mentioned, and reaches not far from the particles with any sensible effect.[17]

The principles of mechanics originated by Galileo and Newton are, as is all of physics today, most easily rendered in terms of particles. Note I am not saying that the "true" objective reality is particles. I have already emphasized that we have no way of knowing what that true reality is. My point here is that the particle model is the easiest way to understand and describe physical phenomena.

In this model, a system of particles can be compounded into a body whose momentum is the vector sum of the momenta of its constituent particles, whose mass is the sum of the masses of the constituents, and whose velocity is the total momentum divided by the total mass (*not* the sum of particle velocities!).

Furthermore, nothing stops us from peering deeper into the nature of "particles" to find out if they can be best described as composite bodies in their own right. This is the reductionism given to us by the ancient atomists, which is the best model of the material world that physics has in the present day.

In 1738, Swiss mathematician Daniel Bernoulli (1700–1782) showed how the pressure of a gas could be understood by assuming the gas is composed of particles colliding with one another and the walls of a container. As we will see, a century later this became known as the *kinetic theory of gases* and, despite intense opposition, constituted one of the earliest scientific triumphs of the atomic model of matter.

I would like to point out an advantage of the particulate view of matter that is not always recognized and exploited. Most people have difficulty understanding the physics that underlies everyday phenomena. However, the observations we make in normal life are easily understood if you think in terms of particle interactions. We can't walk through a wall because the electrons in our bodies electrostatically repel the electrons in the wall. Our hands warm when we rub them together because we are transforming the kinetic energy in the motion of our hands to kinetic energy of the particles in our hands, thereby raising their temperature. An electric current is the flow of electrons from point to point. When we talk, the vibration of our vocal cords causes a pressure wave that passes through the air to set a listener's eardrum vibrating. That pressure wave is composed of a series of regions where the density of air particles is alternately higher and lower that moves from the mouth to the ear.

And, of course, as we will see, light is not some occult force but the passage of particles called photons from a source to a detector such as the eye.

MECHANICAL PHILOSOPHY

Historian David Lindberg describes how Aristotle's physics remained unchallenged in the later Middle Ages until the rival physics of Epicurean atomism became known through the rediscovery of Lucretius's *De rerum natura*. Atomism contributed to the "mechanical philosophy" that, by the end of the seventeenth century, had become dominant. The key figures were Galileo in Italy, Descartes and Gassendi in France, and Boyle and Newton in England, with many others also contributing.

All except Descartes adopted the picture of atoms and the void. Instead, he viewed the universe as a continuum of matter filling all of space and described the motion of planets as being moved by rotating bands of matter called vortices. So while it was not the model of the solar system that survived, Descartes's model was the first attempt to provide a mechanical explanation. And so, the organic universe of medieval metaphysics and cosmology was routed by the lifeless machinery of the new materialists.[18]

The physics of atomism presented no problems for Galileo, Newton, and other theists of the time, nor does it today. They simply reject the cosmological and metaphysical implications. Atomism, as originally presented, posits an infinite universe of eternal, uncreated material particles acted on by impersonal, nonliving, mechanical forces. Most religions imagine a finite, created universe containing not only material particles but also immaterial souls, acted on by personal, vital, supernatural forces.

The Christian atomists of the seventeenth century accepted the particle aspects of atomism, but they rejected the notion that reality is nothing but atoms and the void. They all saw what happened in 1600 to Bruno (see chapter 2), who preached not only that everything was made of atoms but also the atomist doctrine of an infinite universe, although his particular teaching was not atheistic but referred to an infinite God. Gassendi also argued that atoms and God can coexist. Galileo never questioned the spiritual authority of the Church, and his atomism seems to have played no part in his trial for teaching heliocentrism.[19] Newton, although accepting the existence of corpuscles and the void, did not view his own theory of corpuscular motion as complete and talked about God continually acting in moving bodies around to suit his plans. In *Principia*, Newton wrote about the atomists of antiquity:

> They are thus compelled to fall back into all the impieties of the most despicable of all sects, of those who are stupid enough to believe that everything happens by chance, and not through a supremely intelligent Providence; of these men who imagine that matter has always necessarily existed everywhere, that it is infinite and eternal.[20]

PRIMARY AND SECONDARY QUALITIES

Despite Newton's objections to the impiety of atomism, the Newtonian mechanistic scheme implied a distinction between two types of physical properties that was first proposed by Democritus and is inherent to the atomic model. Recall from chapter 1 that Democritus was quoted as saying, "By convention sweet, by convention bitter, by convention hot, by convention cold, by convention color; but in reality atoms and the void."

According to philosopher Lawrence Nolan, as science gradually developed into a field separate from philosophy, it became characterized, especially after Galileo, by the reliance on sensory observation and controlled experimentation as the primary, if not the only, reliable sources of knowledge about the world.[21] Nolan notes that a distinction between *primary* and *secondary* properties was fundamental to the mechanistic model of the universe developed in the seventeenth century.

The new model sought to explain all physical phenomena in terms of the mechanical properties of the small, invisible parts (atoms) that constitute matter. These are primary. They are all that is needed to explain how things work. Secondary properties, such as color or sound, play no role in that explanation. As Nolan puts it, "The color of a clock or the sound it makes when it chimes on the hour are [*sic*] irrelevant to understanding how a clock works." All that matters are the size, shape, and motion of its gears.[22]

A major objection to this view, going back to Aristotle, is that the primary properties are unobservable while the so-called secondary properties are what we actually detect with our own two eyes and other senses. The distinction is perhaps less important today, where we can use our scientific instruments to measure the mass, energy, and other primary properties of particles. The main difference we now recognize is that many secondary properties are, as Democritus noted, "conventions" that we use to describe the subjective reactions we mentally experience as our brains process what our senses detect. For example, I might find an apple tastes sweet, while you find it tastes sour.

However, not all secondary properties are simple, subjective artifacts of the human cognitive system. Recall the discussion in chapter 1 about *wetness*. This is a property of water and other liquids that results from the arrangement of the molecules of the liquid and is not a primary property present in the molecules themselves. Today we call such properties *emergent*. While it is true that wetness is something we sense when we touch water, the property of wetness can be objectively registered with instruments independent of direct involvement of any human sensory apparatus.

So I will make a distinction between secondary *properties* and secondary *qualities* that I have not seen made by philosophers writing on the subject. The physical detectability of wetness is a secondary property that is independent of human involvement, while the conscious sensation of wetness is a secondary quality that involves the human cognitive system.

A long-standing controversy exists among philosophers about the perceiver dependence of the secondary quality color, which

Democritus listed as one of his conventions, along with bitter, sweet, hot, and cold.[23] Whether something tastes bitter or sweet, or feels hot or cold, is clearly subjective, dependent on human sensory perception. Similarly, color, such as the redness of a tomato, is a qualitative experience and so is a secondary quality. But each of these phenomena is also associated with objective properties. Hotness and coldness are related to what you read on a thermometer. The color "red" is the name we apply to the way our brains react when our eyes are hit with photons (atoms of light) in the energy range from 1.8 to 2 electron-volts.

Considering the role of the human cognitive system in describing secondary qualities leads us into a discussion of the brain and the still-controversial question of whether matter alone is sufficient to explain the subjective experiences we have, such as pain, which are called *qualia*. However, this is not a subject for this chapter and will be deferred until chapter 13.

OTHER ATOMISTS

Another important, believing scientist who helped spread the gospel of atomism was chemist Robert Boyle (1627–1691). Boyle's law says that the pressure and volume of a gas are inversely proportional when the gas is at a fixed temperature. Bernoulli was able to derive Boyle's law from the kinetic theory of gasses, as mentioned previously. Boyle made no original contributions to the atomic model itself and, like most of his contemporaries, imbued them with divine purpose.[24] However, his experiments did get people thinking again about the void.

Richard Bentley (1662–1742), chaplain to the bishop of Worcester, was also an ardent atomist and anticipated that the universe was mostly void. But he could not see how purposeless matter could account for the ordered structure of the world.[25]

The noted philosopher John Locke (1632–1704) also supported

atomism but expressed skepticism that theory alone, without observations, can elucidate the fundamental nature of things. However,
recall that atomism was based not just on thought but also on observations. Locke also followed other Christian atomists in rejecting
the notion that "things entirely devoid of knowledge, acting blindly,
could produce a being endowed with awareness."[26]

This view was widespread as science developed further in the
eighteenth century. Indeed, it was a problem that concerned philosophers in Europe and in the Arabic-speaking world during
the Middle Ages. One solution was to imbue atoms with life and
intelligence of their own. A non-supernatural duality was envisaged in which two types of matter existed: organic and inorganic.
The French philosophers Pierre Louis Maupertuis (1698–1759) and
Denis Diderot (1713–1784) promoted this doctrine.[27]

Animate atoms made it possible to eliminate the need for
a divine hand in assembling them into living things, especially
thinking humans. Diderot made this suggestion along with Paul-
Henri Thiry, Baron d'Holbach (1723–1789), both firm atheists at a
time when there were few around. They collaborated in assembling
the colossal thirty-five-volume *Encyclopédie*.[28]

French physician Julien Offray de la Mettrie (1709–1751) provided what was perhaps the first fully materialist, atheist philosophical doctrine. He argued that the human body is a purely material
machine and that there was no soul or afterlife.[29] Like others of his
era, he could not see how chance could produce the world as we
see it. However, he also rejected God as the source and said the
world results from the operation of natural laws that remain to be
discovered.

MORE ANTIATOMISTS

The antiatomists of the seventeenth and eighteenth centuries were
motivated by their Christian beliefs and used religious rather than

scientific arguments to support their positions. These included Descartes, who, although a proponent of mechanics, was deeply wedded to the duality of mind and body and could not believe that everything is reducible to atoms.[30]

Another great philosopher of the period who objected to *physical* atoms was Gottfried Wilhelm Leibniz (1646–1716). In their place, he imagined *metaphysical* atoms called "monads," little immaterial soul-like objects that formed the substance of existence.[31] He never explained how we would be able to demonstrate their existence. But, then, the concept of empirical verification, introduced by Galileo, had not yet taken hold.

4

THE CHEMICAL ATOM

Matter, though divisible in an extreme degree, is
nevertheless not infinitely divisible. That is, there
must be some point beyond which we cannot go in
the division of matter. . . . I have chosen the word
"atom" to signify these ultimate particles.

—John Dalton[1]

FROM ALCHEMY TO CHEMISTRY

For over two thousand years, the atomic theory of matter had no empirical confirmation. Nevertheless, the model of particles moving around in empty space was fundamental to Newtonian mechanics and the scientific revolution that followed. Finally, at the beginning of the nineteenth century, experiments in chemistry began to provide indirect evidence in support of the atomic picture.

Chemistry grew, at least in part, out of the ancient art of alchemy, which was characterized by attempts to transmute base metals into gold or silver. However, alchemy as practiced from the Middle Ages well into the seventeenth century was motivated by more than a desire for instant wealth. Alchemists sought a magical substance called the "philosopher's stone" that they hoped would be the key to transforming imperfect, corruptible substances into that which is in a perfect, incorruptible state. It was thought that the philosopher's stone would cure illness and even generate immortality. It was the "elixir of life."

Alchemy was practiced in ancient Egypt, Greece, India, and China, and flourished in medieval Europe and the Arabic empire. Indeed, the Islamic scholar Jābir ibn Hayān (ca. 815), known in the West as Geber the Alchemist, created much of what we now regard as the laboratory practice of chemistry while he practiced alchemy.[2] Note that the word *alchemy* is a combination of the Arabic *al* (the) and the Greek *chimeia* (chemistry).

Everywhere alchemy was practiced, it was closely tied to the indigenous religious and spiritual systems. It was not science as we know it today. While the Catholic Church generally rejected those occult beliefs not part of its own dogma, alchemy found approval from several popes as well as from Martin Luther (1483–1546) and other Christian leaders. Pope John XXII (1244–1334) was a practicing alchemist. The famous scholar and saint, Albert the Great (ca. 1200–1280), who tried to reconcile science and religion, was a master alchemist. He passed on his knowledge to his student, the even more famous scholar and saint Thomas Aquinas, who wrote several treatises on alchemy.

In England, Henry VIII (1491–1547) and Elizabeth I (1533–1603) actively supported alchemy. Queen Elizabeth provided funding for one of the most prominent alchemists in history, John Dee (1527–1608). Dee was not only an alchemist but also an accomplished mathematician and astronomer whose positive contributions to these fields, notably celestial navigation, were diminished by his obsession with magic.[3] In 1582, Dee claimed an angel gave him a crystal ball that enabled him to communicate with spirits and foretell the future. The actual crystal can be viewed in the British Museum in room 46 of the Tudor collection.[4]

Alchemy was a strange combination of empirical pseudoscience and mysticism. Alchemists had a vast knowledge of how various materials behave when mixed with other materials and heated. One of the favorite materials was mercury, or quicksilver. Since it looked like silver, it was a good candidate to aid in the manufacture of that precious metal. Indeed, if you dissolve mercury in nitric acid and then mix in some lead, a silvery residue appears. However, the residue turns out to be not silver but quicksilver. What happens is the lead reacts more strongly with the nitric acid, and so the mercury leeches back out.

The endeavors of alchemists could certainly be labeled as "experiments." However, these experiments were not performed very systematically, and reported observations were usually couched in codes and arcane language to keep any knowledge gained secret, which is an important difference with real science. Although thousands of volumes of alchemic literature going back centuries were in circulation, virtually all of it was useless gibberish.

Although Isaac Newton is immortalized for his physics, he spent more of his time on and wrote more words about alchemy than physics. In a fascinating book titled *Isaac Newton: The Last Sorcerer*, author Michael White writes:

> Like all European alchemists from the Dark Ages to the beginning of the scientific era and beyond, Newton was motivated by a deep-rooted commitment to the notion that alchemical wisdom extended back to ancient times. The Hermetic tradition—the body of alchemical knowledge—was believed to have originated in the mists of time and to have been given to humanity through supernatural agents.[5]

It never occurred to Newton, the great mechanic and atomist, to take seriously the atomist view of life as being purely the result of the mechanical motions of particles. Newton regarded the philosopher's stone to be the *elixir vitae*, the miraculous substance of life.

Newton's alchemy was far more careful and systematic than most of the alchemists preceding him, as befits his reputation as the greatest scientist of all time. However, he was still a man of his time, with a strong belief in magic and the supernatural. He despised the Catholic Church, agreeing with the radical Puritan view that it is the antichrist, the devil, and the "Whore of Babylon" of the book of Revelation. He had little more use for the Anglican Church, and although he was Lucasian Professor at Trinity College, Cambridge (the position currently held by cosmologist Stephen Hawking), he dismissed the Trinity as inherently illogical and a corrupt contrivance of the Church.[6]

Besides Newton's real science and his alchemy, he spent huge amounts of time trying to extract dates for the fulfilling of the prophecies of the Bible, and alchemic history somehow aided in that task. He predicted that the Second Coming of Christ would be in 1948.[7] I don't think that followed from the laws of motion.

This is not the only place where Newton's spiritual beliefs impacted his science. For example, he believed the patterns of motion of the planets in the solar system could not be explained scientifically and so God continually adjusted these motions.[8]

While Newton was dabbling in the occult, his contemporaries, notably Robert Boyle, were in the process of transforming the magical art of alchemy into the science of chemistry. Thus, Newton truly was the "last sorcerer."

Despite the mystical aspects of alchemy, its experimental practitioners nevertheless accumulated considerable factual knowledge and developed many of the methods that became the modern experimental science of chemistry. As the new science of physics came into being in the seventeenth century, so did the new science of chemistry. Here the key figure was not Newton, who, as we saw, was still mired in the mystical arts, but Boyle. While still doing alchemy himself, in 1661 Boyle published *The Sceptical Chymist*, where he distinguished chemistry as a separate art from alchemy and medicine.

Boyle was very much an atomist and mechanical philosopher. He introduced what we now call *Boyle's law*, which says the pressure and volume of an "ideal" gas are inversely proportional for a gas at a fixed temperature. This would form part of the ideal gas law that, as we will see below, would later be derived from and provide empirical evidence for the atomic model.

THE ELEMENTS

Practical chemistry is an ancient art. After all, metallurgy goes back to prehistoric times. A copper axe from 5500 BCE has been found in Serbia. Many of the other metals—gold, silver, lead, iron, tin—that were mined and smelted in ancient times are now identified as chemical elements. When their ores were heated to very high temperatures, they separated out from the rest of the materials in which they were embedded. However, the elements could not be reduced further—that is, until the twentieth century.

Similarly, irreducible substances such as sulfur, mercury, zinc, arsenic, antimony, and chromium were identified for a total of thirteen elements known prior to the Common Era, although they were not recognized as elementary at the time. By 1800, thirty-four elements had been identified, with another fifty uncovered in the nineteenth century. The twentieth century added twenty-nine more elements, of which sixteen were synthesized in particle accelerators. At this writing, five more have been synthesized in the current century. Norman E. Holden of the National Nuclear Data Center at Brookhaven National Laboratory has provided a history of the elements recognized up to 2004.[9]

Still, prior to the eighteenth century, the ancient belief was widely held that these irreducible materials were not in fact elementary but composed of fire, earth, air, and water. Then, in France, Antoine Lavoisier (1743–1794) showed that water and air were not elementary substances but instead were composed of elements such as hydrogen, oxygen, and nitrogen. He identified oxygen as a component of air and responsible for combustion, disproving the theory that bodies contain a substance called *phlogiston* that is released during combustion.

Making precise quantitative measurements, Lavoisier demonstrated that the total mass of the substances involved in a chemical reaction does not change between the initial and final state, even though the substances themselves may change. This is the *law of conservation of mass* that, more than a century later, Einstein would show must be modified because mass can be created and destroyed from energy. This effect is negligible for chemical reactions, and, after Einstein, mass conservation was simply subsumed in the more general principle of conservation of energy.

By the time the nineteenth century opened, chemistry had expunged itself of alchemy, and laboratory chemists focused on careful analytical methods to build up a storehouse of knowledge on the properties of both elements and the compounds that were formed when they combined. The data eventually led to the periodic table of the chemical elements. Proposed in 1869 by Russian chemist Dmitri Mendeleev, it hangs on the walls of all chemistry classrooms today.

THE CHEMICAL ATOMS

In a lecture before the Royal Institution in 1803, John Dalton (1766–1844) proposed an atomic theory of matter that was really not that much different from that of the ancient atomists:

- All matter is composed of atoms.
- Atoms cannot be made or destroyed.
- All atoms of the same element are identical.
- Different elements have different types of atoms.
- Chemical reactions occur when atoms are rearranged.
- Compounds are formed from atoms of the constituent elements.

However, Dalton did have hard, quantitative evidence—from his own experiments and others'—that provided an empirical basis none of the earlier atomists had available to them. He put these together in *A New System of Chemical Philosophy*, published in 1808.[10] Dalton did not, however, assume that the atoms were necessarily particulate in nature.

In 1806 French chemist Joseph Proust (1754–1826) proposed what is called the *law of definite proportions,* which says that when two elements combine to form a compound, they always do so with the same proportion of mass. For example, when hydrogen and oxygen unite to form water, they always do so in the mass ratio of 1 to 8. Dalton added that the ratio is always a simple proportion of whole numbers, that is, a rational number. This is called the *law of multiple proportions.*

However, Dalton insisted that the ratio was always the same number for the same two elements, while evidence accumulated that frequently two elements combine into more than a single whole-number ratio. He defined the *atomic weight* as the mass of an atom in units of the mass of hydrogen. That is, hydrogen is assigned an atomic weight of 1. Dalton assumed that the water molecule is HO, and this implied that the atomic weight of oxygen is 8. With the discovery of hydrogen peroxide, Dalton's assumption implied that its formula was HO_2. However, it was eventually figured out that water is H_2O ("dihydrogen monoxide"), hydrogen peroxide is H_2O_2, and the atomic weight of oxygen is not 8 but 16.

Today the atomic weight is defined to be exactly 12 for the carbon atom that has a nucleus containing six protons and six neutrons surrounded by six electrons: C^{12}. This definition is unfortunately anthropocentric. We humans just happen to have a lot of carbon in our bodies.

Various *isotopes* of carbon exist with different numbers of neutrons in the nucleus. With this definition, the atomic weight of the hydrogen atom is not exactly 1 but 1.00794. However, 1 is sufficiently accurate for our purposes. Atomic weight is also referred to as atomic mass, molecular weight, or molecular mass. I will simply call it atomic weight, even in the case of a molecule composed of many atoms.

If you look at the periodic table, you will see that the ordering is not by atomic weight. The order number is called the *atomic number* and in constructing the table, Mendeleev moved some elements around so that they also fell into place based on their chemical properties. Thus we have elements with similar properties arranged vertically, with the very chemically active elements H, Li, Na, K, and the like in the first column and the inert elements He, Ne, Ar, Kr, and the like in the nineteenth column. Actually, the older table that you will still see in some classrooms had just eight columns; the modern table has been expanded to nineteen columns, as shown in figure 4.1.[11] It would not be until the chemical atoms were found to have a substructure, which quantum mechanics explained, that the science underlying the periodic table would be understood.

hydrogen 1 H 1.0079																	helium 2 He 4.0026
lithium 3 Li 6.941	beryllium 4 Be 9.0122											boron 5 B 10.811	carbon 6 C 12.011	nitrogen 7 N 14.007	oxygen 8 O 15.999	fluorine 9 F 18.998	neon 10 Ne 20.180
sodium 11 Na 22.990	magnesium 12 Mg 24.305											aluminum 13 Al 26.982	silicon 14 Si 28.086	phosphorus 15 P 30.974	sulfur 16 S 32.065	chlorine 17 Cl 35.453	argon 18 Ar 39.948
potassium 19 K 39.098	calcium 20 Ca 40.078	scandium 21 Sc 44.956	titanium 22 Ti 47.867	vanadium 23 V 50.942	chromium 24 Cr 51.996	manganese 25 Mn 54.938	iron 26 Fe 55.845	cobalt 27 Co 58.933	nickel 28 Ni 58.693	copper 29 Cu 63.546	zinc 30 Zn 65.39	gallium 31 Ga 69.723	germanium 32 Ge 72.61	arsenic 33 As 74.922	selenium 34 Se 78.96	bromine 35 Br 79.904	krypton 36 Kr 83.80
rubidium 37 Rb 85.468	strontium 38 Sr 87.62	yttrium 39 Y 88.906	zirconium 40 Zr 91.224	niobium 41 Nb 92.906	molybdenum 42 Mo 95.94	technetium 43 Tc [98]	ruthenium 44 Ru 101.07	rhodium 45 Rh 102.91	palladium 46 Pd 106.42	silver 47 Ag 107.87	cadmium 48 Cd 112.41	indium 49 In 114.82	tin 50 Sn 118.71	antimony 51 Sb 121.76	tellurium 52 Te 127.60	iodine 53 I 126.90	xenon 54 Xe 131.29
caesium 55 Cs 132.91	barium 56 Ba 137.33	57-70 *	lutetium 71 Lu 174.97	hafnium 72 Hf 178.49	tantalum 73 Ta 180.95	tungsten 74 W 183.84	rhenium 75 Re 186.21	osmium 76 Os 190.23	iridium 77 Ir 192.22	platinum 78 Pt 195.08	gold 79 Au 196.97	mercury 80 Hg 200.59	thallium 81 Tl 204.38	lead 82 Pb 208.98	bismuth 83 Bi 121.76	polonium 84 Po [209]	astatine 85 At [210]	radon 86 Rn [222]
francium 87 Fr [223]	radium 88 Ra [226]	89-102 **	lawrencium 103 Lr [262]	rutherfordium 104 Rf [261]	dubnium 105 Db [262]	seaborgium 106 Sg [266]	bohrium 107 Bh [264]	hassium 108 Hs [269]	meitnerium 109 Mt [268]	ununnilium 110 Uun [271]	unununium 111 Uuu [272]	ununbium 112 Uub [277]		ununquadium 114 Uuq [289]				

*Lanthanide series	lanthanum 57 La 138.91	cerium 58 Ce 140.12	praseodymium 59 Pr 140.91	neodymium 60 Nd 144.24	promethium 61 Pm [145]	samarium 62 Sm 150.36	europium 63 Eu 151.96	gadolinium 64 Gd 157.25	terbium 65 Tb 158.93	dysprosium 66 Dy 162.50	holmium 67 Ho 164.93	erbium 68 Er 167.26	thulium 69 Tm 168.93	ytterbium 70 Yb 173.04
**Actinide series	actinium 89 Ac [227]	thorium 90 Th 232.04	protactinium 91 Pa 231.04	uranium 92 U 238.03	neptunium 93 Np [237]	plutonium 94 Pu [244]	americium 95 Am [243]	curium 96 Cm [247]	berkelium 97 Bk [247]	californium 98 Cf [251]	einsteinium 99 Es [252]	fermium 100 Fm [257]	mendelevium 101 Md [258]	nobelium 102 No [259]

Figure 4.1. The modern periodic table of the chemical elements.

THE CHEMICAL OPPOSITION

That is not to say that the model of atoms and the void was accepted without opposition during the nineteenth century. In fact, the opposition was fierce, especially in France. Many chemists, philosophers, and even a few physicists were far from convinced of the existence of atoms. In an 1836 lecture before the Chemical Society in London, French chemist Jean-Baptiste Dumas (1800–1884) said, "If I were the master, I would outlaw the word 'atom' from science, convinced as I am that it goes far beyond experiments."[12]

The objection of Dumas and his fellow antiatomists of the time was basically that no one had ever seen an atom or a molecule. Henri Sainte-Claire Deville (1818–1881) wrote: "I accept neither Avogadro's law, nor atoms, nor molecules, nor forces, nor particular states of matter; I absolutely refuse to believe in what I cannot see and can even less imagine."[13] Deville didn't have much of an imagination.

Another French chemist named Marcellin Berthelot (1827–1907) was a dedicated antiatomist who also held a government position that enabled him to suppress the teaching of atomism in French schools well into the twentieth century. As late as the 1960s, governmental decrees required that chemistry be taught through "facts only." Authors of textbooks complied by relegating the atomic theory to an afterthought, reminiscent of today's America where many high-school textbooks and teachers still present evolution as "just a theory, not a fact." It is also reminiscent of the Church's order to Galileo to teach heliocentrism as just a theory.

THE PHILOSOPHICAL OPPOSITION

Nineteenth-century theologians had little to say about atomism. The deism and atheism of the seventeenth century had pretty much fizzled after the bloody failure of the French Revolution. The age of reason in Europe was replaced by the romantic movement in literature and art, and by the "Great Awakening" when Protestantism in America turned away from theology and tradition toward emotional and spiritual renewal. The Thomas Jeffersons, Benjamin Franklins, and Denis Diderots were gone, and no one except scientists thought much anymore about science. As today, most scientists preferred it that way. Nineteenth-century atomists were mostly religious believers who stuck to the physics and chemistry and ignored Epicurean atheism. But that did not make Epicurean atheism go away.

Once chemistry was clearly separated from alchemy, it could join physics, which was already fully materialistic as a science that could be pursued independent of one's religious inclinations. Some antiatomists, such as Dumas and Duhem, were fervent Catholics but did not base their arguments on metaphysics.[14]

The same could not be said about biology. Darwinian evolution certainly caught the attention of educated clergy but was not especially noticed by the general public until the early twentieth century. The role of atoms in evolution also had to await further developments, in particular the discovery of the structure of DNA, which did not occur until 1953.

Several philosophers in the early nineteenth century strongly opposed atomism, notably Georg Wilhelm Friedrich Hegel (1770–1831) and Arthur Schopenhauer (1788–1860). Their objections were based on metaphysics as well as skepticism about the physics.

Hegel's metaphysical objections followed from his notion that reality cannot be separated into parts. And to Hegel, atomism was metaphysics:

> Since even today atomism is favored by those physicists who re-fuse to deal in metaphysics, it must be reminded here that one cannot escape metaphysics, or, more specifically, the reduction of nature to thoughts, by throwing oneself into the arms of atomism, since the atom is itself actually a thought, and consequently the apprehension of matter as composed of atoms is a metaphysical apprehension.[15]

As historian Bernard Pullman points out, since its inception the atomic theory has been dogged by a criticism expressed by Aristotle, Cicero, and even Newton that was brought up again by Hegel: How is it that the latent properties of complex compounds form from atoms just bouncing off one another in the void? It would take quantum mechanics to answer that.[16]

Schopenhauer's objections were based on his strange philosophy of "the world as will and representation," of which I could never make any sense.[17] So let me just use his own words:

> Matter and intellect are two interwoven and complementary entities: they exist only for one another and relative to one another. Matter is a representation of the intellect; the intellect is the only thing, and it is in its representation that matter exists. United, they constitute the world as representation, or Kant's phenomenon, in other words, something secondary. The primary thing is that which manifests itself, the thing in itself, in which we shall learn to recognize will.

As for atomism, Schopenhauer is pitiless. He calls it

> a crude materialism, whose self-perception of originality is matched only by its shallowness; disguised as a vital force, which is nothing more than a foolish sham, it pretends to explain manifestations of life by means of physical and chemical forces, to cause them to come from certain mechanical actions of matter, such as position, shape, and motion in space; it purports to reduce all forces in nature to action and reaction.[18]

Despite all these philosophical objections, the physics and chemistry of atoms would emerge triumphant.

5

ATOMS REVEALED

One of the principal objects of theoretical research is to find the point of view from which the subject appears in the greatest simplicity.
                    —Josiah Willard Gibbs[1]

HEAT AND MOTION

As mentioned in chapter 3, in 1738 Daniel Bernoulli proposed that a gas is composed of particles that bounce around and collide with one another and with the walls of the container. Applying Newtonian mechanics, he was able to prove Boyle's law, which said that the pressure of a gas at constant temperature is inversely proportional to its volume.

Other scientists who early anticipated kinetic theory included the polymath scientist and poet Mikhail Lomonosov (1711–1765) in Russia, physicist Georges-Louis Le Sage (1724–1803) in Switzerland, and amateur scientist John Herapath (1790–1868) in England.

Lomonosov, born a poor fisherman's son in Archangel, was an incredible genius—far ahead of his time. He rejected the phlogiston theory of combustion, realized that heat is a form of motion, discovered the atmosphere of Venus, proposed the wave theory of light, and experimentally confirmed the principle of conservation of matter that had confounded Robert Boyle and other better-remembered Western scientists. He presaged William Thomson

(Lord Kelvin, 1824–1907) in the concept of absolute zero. He also showed that Boyle's law would break down at high pressure. In all this, he assumed matter was composed of particles. Lomonosov's story, unfamiliar to most in the West, has been told in a recent article in the magazine *Physics Today*.[2]

Le Sage invented an electric telegraph and proposed a mechanical theory of gravity. While he independently proposed kinetic theory, it made no improvement on that of Bernoulli and was wrong in several ways.

Herapath's paper was rejected by the Royal Society in 1820, and although he managed to have his ideas published elsewhere, they were generally ignored. But he was on the right track.[3]

Despite the largely unrecognized success of Bernoulli's model and these other tentative, independent proposals, the professional physics community centered in western Europe was slow to further develop the atomic model of matter because of its fundamental misunderstanding of the nature of heat. The majority view was that heat was a fluid called *caloric* that flowed from one body to another. However, this was mistaken, and the story of that discovery is a great tale in itself.

As has often happened in the history of physics, military applications provided both an incentive for research and a source of crucial knowledge from that research. During the American Revolutionary War, a New England physicist named Benjamin Thompson (1753–1814) served with the Loyalist forces. After the war, he moved to London and in 1784 was knighted by King George III. From there, Thompson moved to the royal court of Bavaria and was named Count Rumford of the Holy Roman Empire. Rumford is the name of the town in New Hampshire where he had lived prior to the war, now known as Concord.

While supervising the boring of cannon for the Bavarian army, Thompson noticed that as long as he kept boring away, heat was continuously generated. If heat were a fluid contained in the cannon, then at some point it should have been all used up. In a paper to the

Royal Society in 1798, Rumford argued against the caloric theory and proposed a connection between heat and motion. It would be a while before the physics community would accept this connection.

THE HEAT ENGINE

In France, another military engineer, Nicolas Léonard Sadi Carnot (1796–1832), who had served under Napoleon, made the first move toward what would become the science of thermodynamics. Steam engines were just coming into use and it was imperative to understand how they worked. With amazing insight, Carnot proposed an abstract heat engine, now called the *Carnot cycle*, which enabled calculations to be made without the complications of any specific working design.

The Carnot cycle is composed of four processes:

1. Heat from a high temperature reservoir causes a gas to expand at constant temperature (isothermally). The expanding gas is used to drive a piston and do work.
2. The gas is expanded further with no heat in or out (adiabatically) and more work is done.
3. Then the piston moves back (say it is attached to a flywheel), compresses the gas isothermally, and heat is exhausted to the low temperature reservoir.
4. The gas returns to its original state with an adiabatic compression.

The final two steps require work to be done on the gas, but the amount is less than the work done during the two expansions, so a net amount of work is done in the cycle.

An idealized heat engine, not necessarily a Carnot engine, is diagrammed in figure 5.1.

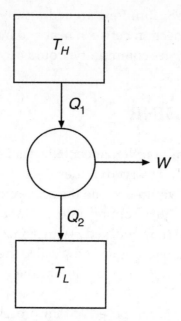

Figure 5.1. A schematic of an idealized heat engine. Heat Q_1 flows from the high-temperature reservoir at temperature T_H to the engine, which does work W and exhausts heat Q_2 to the low-temperature reservoir at temperature T_L. The efficiency of the engine is $\varepsilon = W/Q_1$. If the arrows are reversed, we have a refrigerator or air conditioner.

The efficiency of this idealized engine is $\varepsilon = W/Q_1$, that is, the work done as a fraction of the heat in. Carnot showed that for the Carnot cycle, ε depends on only the temperatures of two heat reservoirs: $\varepsilon = 1 - T_L/T_H$. The higher-temperature reservoir provides the heat, and waste heat is exhausted into the lower-temperature reservoir. In a steam engine, a furnace provides the heat and the environment acts as the exhaust reservoir. Carnot argued that no practical heat engine operating between the same two temperatures could be more efficient than the Carnot cycle.

The Carnot cycle and other idealized engines can be reversed so that work is done on the gas while heat is pumped from the lower-temperature reservoir to the higher-temperature one. In that case, the cycle acts as a refrigerator or air conditioner.

Carnot realized that neither a perfect heat engine, nor a perfect refrigerator or air conditioner, was possible. Consider a perfect air conditioner. That would be one that moves heat from a lower-temperature area, such as a room, to a higher-temperature area, such as the environment outside the room, without requiring any work being done. Such a device could be used to produce a heat engine with 100 percent efficiency, that is, a perpetual-motion machine, by taking the exhaust heat from the engine and pumping it back to its higher-temperature source.

However, heat is never observed to flow from a lower temperature to a higher temperature in the absence of work being done on the system. This principle would later become the simplest statement of the *second law of thermodynamics*. But physicists had still to come up with the *first law*.

CONSERVATION OF ENERGY AND THE FIRST LAW

According to historian Thomas Kuhn, the discovery of the fundamental physics principle of conservation of energy was simultaneous among several individuals.[4] The term *vis viva*, which means "living force" in Latin, is an early expression for energy. It was used by Leibniz to refer to the product of the mass and square of the speed of a body, mv^2. He observed that, for a system of several bodies, the sum $m_1v_1^2 + m_2v_2^2 + m_3v_3^2 + \ldots$ was often conserved. Later it was shown that the kinetic energy of a particle (for speeds much less than the speed of light) is $\frac{1}{2}mv^2$. This was perhaps the first quantitative statement of conservation of energy. He also noticed that $mgh + mv^2$ was conserved, thus introducing what was later shown to be the gravitational potential energy of a body of mass m at a height h above the ground, where g is the acceleration due to gravity.

In 1841, Julius Robert von Mayer (1814–1878) made the first clear statement of what is perhaps the most important principle in physics:

Energy can be neither created nor destroyed.[5]

Mayer received little recognition at the time, although he was eventually awarded the German honorific "von," equivalent to knighthood.

In 1850, Rudolf Clausius (1822–1888) formulated the *first law of thermodynamics*:

In a thermodynamic process, the increment in the internal energy of a system is equal to the difference between the increment of heat accumulated by the system and the increment of work done by it.[6]

This wording is still used today. In the case of the heat engine diagrammed in figure 5.1, $Q_1 = W + Q_2$.

In 1847, Hermann von Helmholtz published a pamphlet titled *Über die Erhaltung der Kraft (On the Conservation of Force)*.[7] Here it must be understood that by "force," Helmholtz was referring to energy. Although rejected by the German journal *Annalen der Physik*, an English translation was published that received considerable notice from English scientists.

The first law of thermodynamics makes the connection between energy, heat, and work explicit. A physical system has some "internal energy," that is yet to be identified. That internal energy can change only if heat is added or subtracted, or if work is done on or by the system. Once again, this is more easily seen with a simple equation:

$$\Delta U = Q - W,$$

where ΔU is the change in internal energy of the system, Q is the heat input, and W is the work done by the system. These are algebraic quantities that can have either sign. Thus, if Q is negative, heat is *output*; if W is negative, work is done *on* the system.

Note that conservation of energy can still be applied in the caloric theory of heat. Caloric is conserved the same way the amount of a fluid, such as water, is conserved when pumped from one point to another. However, caloric is not conserved in a heat engine because mechanical work is done and heat is lost. Recognizing that heat was mechanical and not caloric was the big step that had to be taken in order to understand thermodynamics.

THE MECHANICAL NATURE OF HEAT

Carnot's work on heat engines was not immediately recognized. However, starting in 1834, the new science of thermodynamics was developed more definitively by Benoît Paul Émile Clapeyron (1799–1864) and others.

In England, James Prescott Joule (1818–1889) raised a theological objection to the notion of Carnot and Clapeyron that heat can be lost. In 1843 he wrote:

> I conceive that this theory . . . is opposed to the recognised principles of philosophy because it leads to the conclusion that *vis viva* may be destroyed by an improper disposition of the apparatus: Thus Mr Clapeyron draws the inference that "the temperature of the fire being 1000°C to 2000°C higher than that of the boiler there is an enormous loss of *vis viva* in the passage of the heat from the furnace to the boiler." Believing that the power to destroy belongs to the Creator alone I affirm . . . that any theory which, when carried out, demands the annihilation of force, is necessarily erroneous.[8]

Joule was thinking of conservation of energy, but he was attributing it to deity. He believed that whatever God creates as a basic agent of nature must remain constant for all time.

Although Joule's motivation was at least partly theological, he had the data to demonstrate that heat was not a fluid but a form of energy. In a whole series of experiments, Joule showed how different forms of mechanical and electrical energy generate heat and measured their equivalence. His most significant experiment, which is discussed in most elementary physics textbooks, was published in 1850.[9] An 1869 engraving of his apparatus is given in figure 5.2.[10]

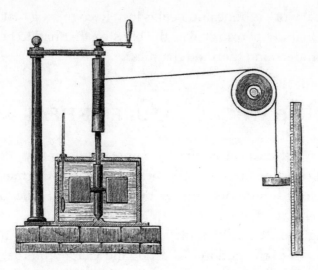

Figure 5.2. Joule's apparatus for measuring the mechanical equivalent of heat. A descending weight attached to a string causes a paddle immersed in water to rotate and the "work" of the falling weight is converted into "heat" by agitating the water and raising its temperature. (Engraving of Joule's apparatus for measuring the mechanical equivalent of heat from *Harper's New Monthly Magazine*, no. 231 [August 1869].)

Shown is a paddle wheel immersed in water that is turned by a falling weight. Multiplying the value of the weight and the distance it falls gives you the work done in stirring the water. A thermometer measures the rise in temperature of the water, from which you can obtain the heat generated. One calorie is the heat needed to raise the temperature of one gram of water one degree Celsius.[11] In the modern International System of Units, the unit of energy and work is named in Joule's honor. He determined the mechanical equivalent of heat to be 4.159 Joules/calorie (the current value is 4.186).

ABSOLUTE ZERO

Another important step in the development of thermodynamics was the notion of absolute temperature, introduced by William Thomson, later made Lord Kelvin, in 1848.[12] Thomson sought an absolute temperature scale that applied for all substances rather than being specific to a particular substance. Carnot's abstract heat engine provided Thomson with the means by which absolute differences in temperature could be estimated by the mechanical effect produced. As he writes,

> The characteristic property of the scale which I now propose is, that all degrees have the same value; that is, that a unit of heat descending from a body A at the temperature $T°$ of this scale, to a body B at the temperature $(T-1)°$, would give out the same mechanical effect, whatever be the number T. This may justly be termed an absolute scale, since its characteristic is quite independent of the physical properties of any specific substance.

Kelvin considered a common thermometer in use at the time, called an *air thermometer*, in which a temperature change is measured by the change in volume of a gas under constant pressure. He then argued that a heat engine operating in reverse (as a refrigerator) could be used to reduce the temperature of a body. However, at some point the volume of the gas in the thermometer would be reduced to zero. Thus, infinite cold can never be reached but some *absolute zero* of temperature must exist. Using experimental data from others, Kelvin estimated this to be –273 degrees Celsius, that is, –273 C. Pretty good. Today we define the absolute temperature in Kelvins (K) to be the Celsius temperature plus 273.15.

THE SECOND LAW OF THERMODYNAMICS

Clausius was also responsible for the first clear statement of the second law of thermodynamics, about which we will have much more to say. Energy conservation does not forbid a perfect refrigerator in which no work is needed to move heat from low temperature to high temperature. But it is an empirical fact that we never see this happening. Heat is always observed to flow from higher to lower temperatures. This is perhaps the simplest way to state the second law of thermodynamics. The other versions just follow. As mentioned previously, if you had a perfect refrigerator, you could use it to cool the low-temperature reservoir of a heat engine below ambient temperature so it could operate with heat from the environment, which would then be serving as the high-temperature reservoir.

Clausius found a way to describe the second law quantitatively. He proposed that there was a quantity that he called *entropy* ("inside transformation" in Greek) that for an isolated system must either stay the same or increase. In the case of two bodies in contact with one another but otherwise isolated, the heat always flows from higher temperature to lower temperature because that is the direction in which the total entropy of the system increases.

KINETIC THEORY

Once the mechanical nature of heat as a form of energy was understood and absolute temperature was defined, the kinetic theory of gases could be developed further. The particulate model provided a simple way to understand physical properties of gases and other fluids described by macroscopic thermodynamics, and the related dynamics of fluids, purely in terms of Newtonian mechanics.

In 1845, the Royal Society rejected a lengthy manuscript submitted by a navigation and gunnery instructor working in Bombay for the East India Company named John James Waterston. Although not correct in all of its details, the paper contained the essentials of kinetic theory, as better-known scientists would develop it a few years later. In particular, Waterston made the important connection between temperature and energy of motion.

One better-known scientist was Rudolf Clausius. In 1857, he published *The Kind of Motion We Call Heat*, in which he describes a gas as composed of particles. Clausius proposed that the absolute temperature of the gas, as anticipated by Waterston, was proportional to the average kinetic energy of the particles. The particles of the gas had translational, rotational, and vibrational energies, and the internal energy was the sum of these.

Although he wasn't the first to propound the kinetic model, Clausius had the skill and reputation needed to spark interest in the subject. In 1860, a brilliant young Scotsman named James Clerk Maxwell (1831–1879) derived an expression for the distribution of speeds of molecules in a gas of a given temperature, although his proof was not airtight. Clausius had assumed all the speeds were equal.

In 1868, an equally brilliant and even younger physicist in Austria named Ludwig Boltzmann (1844–1906) provided a more convincing proof of what became known as the *Maxwell-Boltzmann distribution*, although the two never met. Boltzmann would be the primary champion of the atomic theory of matter and provide its theoretical foundation based on statistical mechanics against the almost-uniform skepticism of others, including Maxwell, and downright harsh opposition from some, notably Ernst Mach.[13]

However, before we get to that, let us consider an example of how kinetic theory is applied. The simplest application of kinetic theory is to what is called an *ideal gas*, which is circularly (but not illogically) defined as any gas that obeys the ideal gas equation.

Let P be the pressure, V be the volume, and T be the absolute temperature of a container of gas. The ideal gas law, formulated by Clapeyron, combines Boyle's law and Gay-Lussac's law, which is better known as Charles's law, and was proposed by Joseph Louis Gay-Lussac (1778–1850). For a fixed mass of gas:

Boyle's law	PV = constant for constant T
Charles's law	V/T = constant for constant P
Ideal gas law	PV/T = constant

In freshman physics classes today, a model in which an ideal gas is just a bunch of billiard-ball-type particles bouncing around inside a container is used to prove that

$$PV = NkT,$$

where N is the number of particles in the container and k is *Boltzmann's constant* (which is just a conversion factor between energy and temperature). This is the precise form of the ideal gas law.

HOW BIG ARE ATOMS?

A key issue in the atomic theory of matter is the scale of atoms. The first step in the direction of establishing the size of atoms was made by the Italian physicist Lorenzo Romano Amedeo Carlo Avogadro, Count of Quaregna and Cerreto (1776–1856). Avogadro proposed what is now called *Avogadro's law,* which states that equal volumes of all gases at the same conditions of temperature and pressure contain the same number of molecules.

At the time, the terms *atoms* and *molecules* were used interchangeably. Avogadro recognized that there were two types of submicroscopic ingredients in matter, the first being elementary molecules, which we now call atoms (or, in my designation, "chemical atoms," which are identified with the chemical elements), and the second being assemblages of atoms, which today we simply call molecules.[14]

Avogadro's number (or Avogadro's constant) is the number of molecules in one *mole* of a gas. The number of moles in a sample of matter is the total mass of the sample divided by the atomic weight of the element or compound that makes up the sample. Allow me to use some simple mathematics that will be easier to understand than anything I can say with words alone. Let's make the following definitions:

N_A Avogadro's number

N the number of molecules (or atoms) in a sample of matter

M the mass of that sample in grams

n the number of moles in that sample

A the atomic weight of the element or compound in the sample

m the actual mass of the atom or molecule in grams

Consider hydrogen, where we make the approximation $A = 1$ (actual value 1.00794). Let the mass of the hydrogen atom be $m = m_H$ in grams. Then,

$$n = \frac{M}{A} = M$$

$$N_A = \frac{N}{n} = \frac{N}{M}$$

$$M = Nm_H$$

$$N_A = \frac{1}{m_H}$$

That is, Avogadro's number, to an approximation sufficient for our purposes, is simply the reciprocal of the mass of a hydrogen atom in grams.

Avogadro did not know the value of N_A. The first estimate, which was reasonably close to the more accurate value later established, was made in 1865 by an Austrian scientist named Johann Josef Loschmidt (1821–1895). He actually estimated another number now called *Loschmidt's number*, but it was essentially the same quantity.

Using the current best value, $N_A = 6.022 \times 10^{23}$, we get $m_H = 1.66 \times 10^{-24}$ grams.

To find the number of molecules in M grams of any substance, simply multiply M by N_A / A. Since the density of water is 1 gram per cubic centimeter (by definition) and its molecular weight is $A = 18$, a cubic centimeter of water contains $N = 3.3 \times 10^{22}$ water molecules.

Until scientists had an estimate for Avogadro's number, they had no idea that the number of molecules in matter was so huge and that the molecules themselves had such small masses. Many chemists thought atoms were balls rubbing against each other, like oranges in a basket. In the next chapter, we will see how the substructure of the chemical atoms was revealed in nuclear physics.

STATISTICAL MECHANICS

Kinetic theory is the simplest example of the application of *statistical mechanics*, developed by Clausius, Maxwell, Max Planck (1858–1947), and Josiah Willard Gibbs (1839–1903), but whose greatest architect was Ludwig Boltzmann. By assuming the atomic picture of matter for all situations, not just an ideal gas, statistical mechanics uses the basic principles of particle mechanics and laws of probability to derive all the principles of macroscopic thermodynamics and fluid mechanics. With statistical mechanics, atoms and the void came home to stay.

Boltzmann had a difficult time convincing others, including Maxwell, that statistical mechanics was a legitimate physical theory. Boltzmann had been a good friend of fellow Austrian Loschmidt, who, as we have seen, had made an estimate of the size of atoms. So Boltzmann knew that the volumes of gas dealt with in the laboratory contained many trillions of atoms. Obviously, it was impossible to keep track of the motion of each, so he was necessarily led to statistical methods. However, at the time, statistics was not to be found in the physicist's toolbox.

Probability and statistics was already a mathematical discipline, thanks mainly to the efforts of Blaise Pascal (1623–1662) and Carl Friedrich Gauss (1777–1855). It was mostly used to calculate betting odds, although there is no record of Pascal using it to determine the odds for *Pascal's wager*.[15]

In any case, the mathematics for calculating probabilities was available to Boltzmann. No one before Boltzmann had ever thought to apply statistics to physics, where the laws were assumed to be certain and not subject to chance. Even Maxwell did not immediately grasp that the Maxwell-Boltzmann distribution, which assumes random motion, was just the probability for a molecule having a speed of a certain value within a certain range. Boltzmann did not appreciate that at first, either. But he nevertheless initiated a revolution in physics thinking that reverberates to this very day.

Boltzmann's monumental work occurred in 1872, when he was twenty-eight years old. He derived a theorem, which was called the *H-theorem* by an English physicist who mistook Boltzmann's German script uppercase *E* for an *H*. This theorem was essentially a proof of the second law of thermodynamics.

In Boltzmann's H-theorem, a large group of randomly moving particles will tend to reach a state where a certain quantity *H* is minimum. Boltzmann identified *H* with the negative of the entropy of the system, and so the H-theorem said that the entropy tends to a maximum, which is just what is implied by the second law of thermodynamics.

Boltzmann's equation for entropy is engraved on his gravestone:

$$S = k\log W,$$

where log is the natural logarithm (the inverse of the exponential function), usually written today as $\log_e$ or ln, and *W* is the number of accessible states of the system.

What exactly do we mean by the "number of accessible states"? Basically, it is the number of *microstates* needed to give a particular *macrostate*. Let me give a simple example. Consider the toss of a pair of dice. Each possible outcome of the toss constitutes a separate accessible microstate of the system of two dice. Now, suppose the "energy" of the system is defined as the sum of the number of dots that turn faceup. Each energy is a different macrostate. The lowest energy is 2, which can occur only one way, "snake eyes," 1 and 1. So the system has one accessible microstate of energy 2. Similarly, the highest energy of 12 has only one accessible microstate, "boxcars," 6 and 6.

Now, energy 3 can occur two ways, so there are two accessible states of that energy. Energy 7 has the most accessible states: 6 and 1, 1 and 6, 5 and 2, 2 and 5, 4 and 3, 3 and 4, for a total of six states. And so on. Figure 5.3 shows the number of accessible microstates for each possible value of energy.

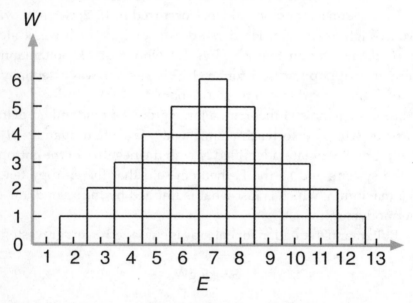

Figure 5.3. The number of microstates that are accessible to a pair of dice for each possible "energy" macrostate (sum of dots).

To get the entropy of the dice as a function of energy, just pick off W from the graph and calculate it from the previous equation. Recall that Boltzmann's constant, k, just coverts temperature units to energy units. Let $k = 1$, in which case entropy is dimensionless. So, for example, for $E = 10$, $W = 3$ and $S = \log(3) = 1.09$.

Note that the lower the value of W—that is, the lower the entropy—the more information we have about the microstates of the system of a given energy. When $E = 7$, the system can be in any of six possible microstates, which is the lowest information state. When $E = 1$ or 12, we have only one possible microstate and maximum information. Communications engineer Claude Shannon made the connection between entropy and information in 1948, thus founding the now vitally important field of information theory that is fundamental to communications and computer science.[16]

This example also makes clear the statistical nature of entropy. The probability of obtaining any particular energy with a toss of the dice is W divided by the total area of the graph, which is 36 for our particular example. Thus, the probability of $E = 7$ (tossing a 7) is $6/36 = 1/6$. The probability of $E = 4$ is $3/36 = 1/12$.

Here the number of samples is small, so we have a wide probability distribution. As the number of samples increases, the probability distribution becomes narrower and narrower so that the outcome becomes more and more predictable. This is known as the *law of large numbers*, or the *central limit theorem* in probability theory.

Most physicists at the time were not convinced by Boltzmann's H-theorem. His friend Loschmidt posed what is called the *irreversibility paradox*. The laws of mechanics are time reversible, so if a set of particle interactions results in H decreasing, the reversed processes will result in H increasing. Maxwell had made a similar argument with a device called "Maxwell's demon," an imaginary creature that can redirect particles so that they flow from cold to hot.

It took Boltzmann a while to realize that his theorem was not a hard-and-fast rule but rather a matter of probabilities. It was not that the system would reach minimum H and stay there forever, as it seems he originally thought, but it can fluctuate away from it. This is what made the H-theorem so hard to accept for great thinkers like Maxwell and Thomson. The second law is just a statistical law. Who had ever heard of a law of physics that was just based on probabilities? What took everyone, including Boltzmann, a long time to realize was that entropy does not always have to increase. It can decrease.

However, all the familiar, macroscopic applications of entropy deal with systems of great numbers of particles, where the probability of entropy decreasing is infinitesimally small. So the second law of thermodynamics is generally considered inviolate. For example, in his 1971 book, *Bioenergetics; the Molecular Basis of Biological Energy Transformations*, Albert L. Lehninger has a diagram showing several processes such as the flow of heat between regions of different temperature and the movement of molecules from regions of different pressure or different concentrations. In the caption he says, "Such flows *never* reverse spontaneously."[17]

> *What never?*
> *No, never.*
> *What never?*
> *Well, hardly ever.*[18]

Consider the flow of heat between two bodies in contact with one another. Picture what happens microscopically, which I have emphasized is always the best way to view any physical process to really understand what is going on. In figure 5.4, the body with black particles is at lower temperature, as indicated by the shorter arrows, on average, that represent their velocity vectors. The body with the white particles is higher temperature, indicated by longer arrows. Energy is transferred between the two bodies by collisions in the separating wall. On average, more energy is transferred from the higher-temperature body to the lower-temperature one because, in a collision between two bodies, the one with the higher energy will transfer more of its energy to the other body. However, when the number of particles is small, there is some chance that occasionally energy will be transferred in the opposite direction, in apparent violation of the second law.

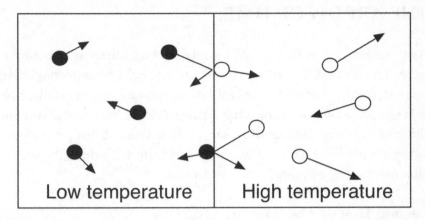

Figure 5.4. The exchange of heat energy between two bodies by collisions with particles in the adjoining wall. The body on the right has the higher temperature. So, on average, the transfer of energy is from right to left, as implied by the second law. However, for small numbers of particles, an occasional transfer from lower temperature to the higher can occur.

Now, in principle, this can happen no matter how many particles are involved. Indeed, all the so-called *irreversible* processes of thermodynamics are, in principle, reversible. Open the door of a closed room and all the air in the room can rush out the door, leaving behind a vacuum. All that has to happen is that at the instant you open the door, all the air molecules need to be moving in that direction. However, this is very, very unlikely because of the huge number of molecules involved. If a room contained just three molecules, no one would be surprised to see them all exit the room when the door was opened.

Thus, Loschmidt's paradox is no paradox. Regardless of the number of particles, every process is, in principle, reversible, exactly as predicted by Newtonian mechanics. This is also true in modern quantum theories.

THE ARROW OF TIME

As Boltzmann eventually realized, there is no inherent direction to time. The second law of thermodynamics, which some physicists have termed the most important law of physics, is not a law at all. It is simply a definition of what Arthur Eddington in 1927 dubbed *the arrow of time*. Rather than saying that the entropy of a closed system must increase, or at best stay constant, with time, the proper statements of the second law is as follows.

Second Law of Thermodynamics

> The arrow of time is specified by the direction in which the total entropy of the universe increases.

As we will see in chapter 10, in the twentieth century, it was found that not all elementary particle processes run in both time directions with equal probability. Some authors have trumpeted this as the source of a fundamental arrow of time. However, this is incorrect. The processes observed still are reversible, with the differences in probability being only on the order of one in a thousand. They are time *asymmetric* but not time *irreversible*. The arrow of time defined by the second law of thermodynamics is of little use when the number of particles involved is small, which is why you never see it used in particle physics.

THE ENERGETIC OPPOSITION

Perhaps the best-known of the late nineteenth- and early twentieth-century antiatomists was the Austrian physicist and philosopher Ernst Mach. Along with chemists Wilhelm Ostwald (1853–1932) and George Helm (1851–1923) in Germany, and physicist/historian Pierre Duhem (mentioned in chapter 3) in France, Mach advocated a physical theory called *energeticism*.[19] This theory was based on the hypothesis that everything is composed of continuous energy. In this scheme, there are no material particles. Thermodynamics is the principle science of the physical world and energy is the governing agent.[20] The goal of science, then, is to "describe and codify the conditions under which matter reacts without having to make any hypothesis whatsoever about the nature of matter itself."[21]

We have seen how, with the rise of the Industrial Revolution in the nineteenth century, the science of thermodynamics developed as a means of describing phenomena involving heat, especially heat engines. Thermodynamics became a remarkably advanced science with enormous practical value.

The nineteenth-century science of thermodynamics was based solely on macroscopic observations and made no assumptions about the structure of matter. Furthermore, it was very sophisticated mathematically. It contained great, general principles such as the zeroth, first, second, and third laws of thermodynamics. Indeed, energy was the key variable that could appear as the heat into or out of a system, the mechanical work done on or by a system, or the internal energy contained inside the system itself. A remarkable set of equations was developed relating the various measurable variables of system such as pressure, temperature, and volume, to other abstract variables such as entropy, enthalpy, and free energy.

At the same time, the atomic theory of matter suffered from the problems we have already discussed, in particular, as Mach especially continued to emphasize, that no one had ever seen an atom. He insisted that science should only concern itself with the observable.

The figure who did the most to promote the general theory of macroscopic thermodynamics was a reclusive American physicist, Josiah Willard Gibbs (1839–1903), with a series of publications in the *Transactions of the Connecticut Academy of Sciences* in 1873, 1876, and 1878. While this was not, to say the obvious, a widely read journal, Gibbs sent copies to all the notable physicists of the time, even translating them into German. Ostwald seized on Gibbs's work as support for energeticism, but Gibbs wasn't in that camp. He was not against atoms; he just did not need them to develop the mathematics of thermodynamics.[22] Indeed, his work provided a foundation for statistical mechanics as well as for physical chemistry, and a year before his death in 1903, he wrote a definitive textbook, *Elementary Statistical Mechanics*.

In any event, the energeticists had good reasons to view thermodynamics as the foundation of physics. Ostwald even denied the existence of matter outright. All was energy. He saw advantages in science adopting energeticism:

> First, natural science would be freed from any hypothesis. Next, there would no longer be any need to be concerned with forces, the existence of which cannot be demonstrated, acting on atoms that cannot be seen. Only quantities of energy involved in the relevant phenomena would matter. Those we can measure, and all we need to know can be expressed in those terms.[23]

Ostwald gained great prominence in chemistry, including an eventual Nobel Prize in 1909. He founded what is called *physical chemistry*, which combines chemistry with thermodynamics in order to understand the role of energy exchanges in chemical reactions. Boltzmann had met Ostwald when the younger man was still getting started, and they kept up a friendly interaction—at least for a while.

However, Ostwald was influenced by Mach and, like many chemists, did not see the particulate theory as particularly convincing or useful. The matter came to a head on September 16, 1895, when Boltzmann and mathematician Felix Klein (1849–1925) debated Ostwald and Helm in front of a large audience in Lübeck. Ostwald was the far better debater, but Boltzmann easily won the day by arguments alone (which shows it can be done).

The chemists tried to argue that all the laws of mechanics

and thermodynamics follow from conservation of energy alone. Boltzmann pointed out that Newtonian mechanics was more than energy conservation, and thermodynamics was more than the first law. In particular, as we have seen, the second law had to be introduced precisely because the first law did not forbid certain processes that are observed not to occur.

The famous Swedish chemist Svante Arrhenius (1859–1927), who attended the debate, wrote later, "The energeticists were thoroughly defeated at every point, above all by Boltzmann, who brilliantly expounded the elements of kinetic theory." Also in attendance was the young Arnold Sommerfeld (1868–1951), who would become a major figure in the new quantum mechanics of the twentieth century. He wrote, "The arguments of Boltzmann broke through. At the time, we mathematicians all stood on Boltzmann's side."[24]

THE POSITIVIST OPPOSITION

Mach's opposition was not so much based on his support for energeticism as his commitment to a philosophy that was fashionable at the time called *positivism*, which was put forward by French philosopher Auguste Comte (1798–1857). Comte, who was mainly interested in applying science to society and is regarded as the founder of sociology, believed that science must "restrict itself to the study of immutable relations that effectively constitute the laws of all observable events." He dismissed any hope to gain "insight into the intimate nature of any entity or into the fundamental way phenomena are produced."[25] Comte summarizes his position:

> The sole purpose of science is, then, to establish phenomenological laws, that is to say, constant relationships between measurable quantities. Any knowledge about the nature of their substratum remains locked in the province of metaphysics.[26]

Mach insisted that the atomic theory was not science, that physicists should stick to describing what they measure—temperature, pressure, heat—and not deal with unobservables. Atoms are, at best, a useful fiction.

Mach's name is well known because of the definition of the *Mach number* as the ratio of a speed to the speed of sound, and he make major contributions to the theory of sound and supersonic motion. He is also recognized for *Mach's principle*, a vague notion that has many different forms but basically says that a body all alone in the universe would have no inertia, so inertia must be the consequence of the other matter in the universe. Einstein made Mach's principle famous by referring to it in developing his general theory of relativity, although he does not seem to have used it anyplace in his derivations. Furthermore, the principle has never been placed on a firm philosophical or mathematical foundation.

After producing poorly received physics books titled *The Conservation of Energy* (1872) and *The Science of Mechanics* (1883), Mach turned increasingly toward writing and lecturing on philosophy, meeting there with greater success. In the 1890s, he crossed paths with Boltzmann as they both settled down in Vienna.

In Vienna, Boltzmann found himself decidedly in the minority as students and the general public flocked to Mach's lectures while few were interested in the demanding work of mathematical physics. Mach continued to assert that there is no way to prove that atoms are objectively real. Before the Viennese Academy of Sciences, Mach stated unequivocally, "I don't believe atoms exist!"[27]

The dispute between Mach and Boltzmann was clearly not over physics but over the philosophy of science. What is it that science tells us about reality? This question is not settled today. Mach's positivism continued into the early twentieth century as the philosophical doctrine of *logical positivism* that was also centered in Vienna with a group called the Vienna Circle. The logical positivists were led by Moritz Schlick and included many of the top philosophers of the period, such as Otto Neurath, Rudolf Carnap, A. J. Ayer, and Hans Reichenbach.

The goal of the logical positivists was to apply formal logic to empiricism. In their view, theology and metaphysics, neither of which had any empirical basis, were meaningless. Mathematics and logic were tautologies. Only the observable was verifiable knowledge. So far, so good.

While logical positivism held the center stage of philosophy for a while, it eventually fell to the recognition of its own inconsistency. How does one verify verifiability? Later, eminent philosophers such as Karl Popper, Hilary Putnam, Willard Van Orman Quine, and Thomas Kuhn pointed out that all observations have a theoretical element to them, they are what is called *theory-laden*. Time is what is measured on a clock, but it is also a quantity, t, appearing in theoretical models. While logical positivism is no longer in fashion, the essential point that we have no way of knowing about anything except by observation remains an accepted principle in the philosophy of science.

Getting back to atoms, are atoms real or just theoretical constructs? Boltzmann was not dogmatic about it. He recognized that "true reality" can never be determined and that science can only hope to apprehend it step-by-step in a series of approximations.[28] Mach's insistence that atoms were inappropriate elements of a theory because they are not observed would be proven grossly wrong when atoms were convincingly observed. Today we have quarks that not only have never been observed, but they are also part of a theory that says they never will be observed. Indeed, if quarks are someday observed, the theory that introduced them in the first place will be falsified!

Unhappy in Vienna, where he felt unappreciated and was becoming increasingly emotionally unhinged, in 1900, Boltzmann moved to Leipzig where Ostwald magnanimously had found him a place. However, he accomplished nothing there and returned to Vienna in 1902. His lectures were well received, but he showed no interest in the new physics that was blossoming with the turn of the century: Röntgen rays, Becquerel rays, and Planck quanta.

Instead, Boltzmann took to lecturing in philosophy, taking over a course taught by Mach, who had retired after suffering a stroke. Boltzmann had difficulty grasping the subject. After attempting to read Hegel, he complained, "what an unclear, senseless torrent of words I was to find there." In 1905 Boltzmann wrote to philosopher Franz Brentano (1838–1917), "Shouldn't the irresistible urge to philosophize be compared to the vomiting caused by migraines, in that something is trying to struggle out even though there is nothing inside?"[29]

In 1904, Boltzmann attacked Schopenhauer in a lecture he gave before the Vienna Philosophical Society that was originally to be titled, "Proof That Schopenhauer Is a Stupid, Ignorant Philosophaster, Scribbling Nonsense and Dispensing Hollow Verbiage That Fundamentally and Forever Rots People's Brains." Actually, Schopenhauer had written these precise words to attack Hegel.[30] Boltzmann changed the title but still came down hard on Schopenhauer.

During these last years, Boltzmann traveled extensively, including two trips to America. In 1904, he attended a scientific meeting in conjunction with the St. Louis World's Fair. In 1905, he gave thirty lectures in broken English at Berkeley. (He could have lectured in German, which was familiar to scholars of the day.) At the time, Berkeley, of all places, was a dry town, and Boltzmann had to steal to a shop in Oakland regularly to get bottles of wine. Passing through dry states on the train back to New York, he had to bribe attendants to bring him wine.[31]

Back in Vienna, Boltzmann found it impossible to continue his lecturing and was increasingly depressed. He did not seem to be aware of the fact that in 1905 Einstein had provided the necessary theory that would ultimately lead to a convincing verification of the atomic picture, as will be described in the next section. In late summer of 1906, Boltzmann traveled with his family to the village of Duino on the Adriatic coast for a respite by the sea. There, on September 5, his fifteen-year-old daughter, Elsa, found him hanging in their hotel room. His funeral in Vienna a few days later was attended by only two physicists. The academic year had not yet begun.

EVIDENCE

Despite the great success of the atomic theory of matter in accounting for the observations made during the great experimental advances of nineteenth-century chemistry and physics, that century still provided no direct empirical support that atoms really existed. But that would finally come early in the twentieth century.

Einstein is remembered for his special and general theories of relativity. But in the same year that he produced special relativity, 1905, Einstein also published two other papers that helped establish once and for all that all material phenomena can be described in terms of elementary particles.

One of these publications proposed that light is particulate in nature. I will forego that discussion until the next chapter. Here, let us take a look at his least-remembered work, which presented a calculation showing how measurements of Brownian motion could be used to determine Avogadro's number with much greater confidence than existed for earlier estimates and, thus, set the scale of atoms.

Recall from chapter 1 that Lucretius wrote how the dust motes seen in a sunbeam dance about "by means of blows unseen." Now, these dust motes are moved about by air currents as well as by random motion. Such motions are difficult to quantify. However, in 1827, Scottish botanist Robert Brown (1773–1858) was examining under a microscope grains of pollen suspended in water. He noted that they moved about randomly and any water-current effects were negligible. This is called *Brownian motion*. Earlier, in 1785, Dutch biologist Jan Ingenhousz (1730–1799) observed the same random motion for coal dust particles in alcohol.

As suggested by Lucretius, the Brownian particles, which are large enough to be seen with a microscope, are bombarded randomly by the unseen molecules of the fluid within which they are suspended. A nice animation of the effect is available on the Internet.[32]

Einstein argued that the more massive the bombarding atoms in the fluid, the greater would be the fluctuation in the pollen particle's motion. Thus, a measure of that fluctuation can be used to determine the mass of the atoms and thus Avogadro's number.[33]

Shortly after Einstein's publication, French physicist Jean Baptiste Perrin carried out experiments on Brownian motion, using Einstein's calculation to determine Avogadro's number. Perrin applied other techniques as well and was very much involved in ultimately achieving a final consensus, among chemists as well as physicists, on the discrete nature of matter and the small size of atoms. Earlier, in 1895, he had shown that cathode rays were particles with negative electrical charge that were identified two years later by J. J. Thomson as "corpuscles." Thomson estimated the mass of the corpuscle to be a thousand times lighter than an atom. In 1894, George Stoney named the corpuscles *electrons*. Today we still regard the electron as a point-like elementary particle.

By the 1920s, the nineteenth-century antiatomists had died off along with their antiatomism. Ostwald finally had been won over by the data and, in 1908, had stated his belief in atoms in a new introduction to his standard chemistry textbook, *Outline of General Chemistry*.

Mach continued to drag his feet, however, writing against atomism as late as 1915, losing the respect of people like Planck and Einstein who had previously held him in high regard. Einstein had (temporarily) adopted a positivist view in defining time as what one reads on a clock, which should have pleased Mach. Mach did not like relativity, however. It was too theoretical. He died in 1916 at the age of seventy-eight.[34]

6

LIGHT AND
THE AETHER

*What, then, is light according to the electromag-
netic theory? It consists of alternate and opposite
rapidly recurring transverse magnetic disturbances,
accompanied with electric displacements, the direction of the electric displacement being at the right
angles to the magnetic disturbance, and both at
right angles to the direction of the ray.*
— **James Clerk Maxwell**[1]

THE NATURE OF LIGHT

Most believing scientists today have no difficulty accepting
the notion that matter is composed of particles. They just
object to the notion that particles are all there is. Before addressing
that issue, however, we need to spend a few chapters tracing the
developments that led to our current comprehension of the nature
of matter. By the time we are finished, we will see how little room
there is left for anything else.

Until the early twentieth century, it was generally assumed
that matter and light were two separate aspects of physical reality.
Light always carried with it a hint of the occult or spiritual. Even
today, people often think of it as "pure energy," not appreciating the

meaning of $E = mc^2$, that mass and rest energy are equivalent. Light is just as much material as a rock.

To talk about light, we must begin with the sense of vision. The ancients held two opposing views on vision. In one, called *extra-mission*, associated with Euclid (ca. 300 BCE) and Claudius Ptolemy (ca. 168 CE), rays of light are emitted by the eye onto an object, where they are then reflected back, as in modern radar, providing information about the object. In the second view, called *intromission*, held by both Aristotle and the atomists, objects themselves emit light that was then detected by the eye. The Arabic astronomer Ibn al-Haytham (ca. 1040), also known as Alhacen, argued convincingly against extramission by pointing out that bright objects can injure the eye, which would not happen if the source of light were the eye itself. He also noted that we are able to see objects at great distances, such as stars in the heavens, so it is unlikely that light from our eyes reaches out throughout the universe.[2]

It was not until the sixteenth century, however, that serious physical models for the nature of light were developed. Pierre Gassendi proposed a particle theory of light that was published in the 1660s after his death. Newton followed in 1675 with his corpuscular theory of light. This conflicted with the picture of light as a wave phenomenon proposed by Robert Hooke in the 1660s and by Christiaan Huygens in 1678.

In the 1660s, Newton had done a series of experiments with prisms that demonstrated a previously unrecognized fact that white light can be broken down into the colors of a rainbow. However, he kept that knowledge secret until 1672, when he was cajoled into sending a lengthy letter to the Royal Society on what became known as the "theory of light and colours."[3]

There he ran into conflict with the man who would become his bitterest enemy over the years, Robert Hooke. Hooke was a very talented instrumentalist and original thinker, but he was nowhere near Newton in intellectual capacity. Each had high opinions of his own superiority, and they clashed incessantly until Hooke's death.

Hooke was the curator of experiments for the Royal Society when Newton submitted his paper and so was assigned the task of analyzing it. He gave it only a cursory look and was unconvinced. He had his own theory of light and wasn't about to let Newton displace it. He wrote:

> For all the experiments & observations I have hitherto made, nay and even those very experiments which he alleged, do seem to me to prove that light is nothing but a pulse or motion propagated through an homogeneous, uniform and transparent medium.[4]

That is, light is a wave and Newton had proposed in the letter that light is a particle.

At first Newton ignored Hooke's claims, but he could not ignore the reaction of the Dutch physicist and astronomer Christiaan Huygens (1629–1695). Newton had far greater respect for Huygens, as did most of the scientific community, than he did for Hooke. At first, Huygens had expressed admiration for Newton's theory. But then, in a series of letters to the Royal Society in 1672 and 1673, Huygens objected that Newton's theory of colors was no theory at all but a hypothesis.[5]

Author Michael White makes an interesting observation:

> To the modern mind, it seems odd that this same attack should be made without any form of experimental back-up from either dissenter. Huygens, like Hooke, could not accept what was then a completely novel approach—that a hypothesis is tested by experiment and dismissed only if the experiment shows that it is wrong. By 1673, Huygens had not conducted a single experiment in an attempt to prove or disprove Newton's theory. Instead he based his response on *a priori* reasoning alone.[6]

The empirical method has a long history. It had been championed by Francis Bacon and implemented by Galileo. However, it was not until Newton that the remnants of Aristotelian thinking were finally buried and experiment was given precedence over theory.

All this led to further vitriolic attacks back and forth between Newton and Hooke. The famous Newtonian line, "If I have seen further it is by standing on ye shoulders of giants" appeared in a 1676 personal letter from Newton to Hooke.[7] Instead of being a grand gesture of humility, as it is usually regarded, White interprets this as a vicious slap at Hooke, who was "so stooped and physically deformed that he had the appearance of a dwarf."[8]

If you eliminate the personal animosity between Newton and Hooke, and the lack of a full appreciation by Huygens and others at the time of the distinction between theoretical hypotheses and experimental facts, the dispute comes down to one over rival theories on the nature of light. Hooke and Huygens, for good reasons, promoted the wave theory while Newton, also for good reasons, promoted the corpuscular theory. In 1672, Newton clarified his position in a paper read before the Royal Society, with Hooke in attendance. Newton admitted that he had argued for the corpuscular nature of light but said, "I do it without any absolute positiveness, as the word *perhaps* intimates, & make it at most but a very plausible consequence of the doctrine, and not a fundamental supposition."[9]

So let us get back to the physics. The wave theory implied that light should exhibit a phenomenon observed for water and sound waves, namely, *diffraction*, the ability to turn corners. Newton objected to the wave theory of light, claiming it did not diffract. We can hear—but we can't see—around corners. Here the great physicist known for his experimental as well as theoretical brilliance failed to empirically test his belief. The diffraction of light is easily observed by holding a card with a small pinhole in it up to a lamp (don't use the sun). Instead of a sharp outline of the hole, you will see a diffuse spot of light. That's diffraction. We *can* see around corners—just not very far.

Hooke had done nothing to empirically test the wave theory. However, Huygens was eventually able to demonstrate how diffraction and other optical effects such as refraction and interference are wave phenomena. He proposed that each point on a wave front acts as the source of a spherical "wavelet." These wavelets then combine to give the next wave front.

So, I would say that at the time of Newton and Huygens, the empirical evidence for the wave theory of light was already solid, while the corpuscular theory had no way of accounting for the observed properties of diffraction and interference. Still, by virtue of his immense authority, Newton's corpuscular theory of light prevailed until 1800 when Thomas Young (1773–1829) performed experiments on light interference that finally won over the scientific community.

By the nineteenth century, then, the wave nature of light was confirmed. This would seem to discredit the basic tenet of atomism, that everything is just particles and void. But the story was not yet over.

Light waves had to be vibrations of some sort of medium, just as sound waves are vibrations of materials such as air, water, and even solids. In the case of light, the medium of vibration was termed the *aether* and was viewed as pervading all of space. All that was needed now was some sort of empirical confirmation of the existence of the aether.

THE AETHER

Two opposing views of the physical world mark the contrast between atomism and antiatomism. Atomism envisages a universe that divides up into material parts separated by empty space—atoms and the void. Antiatomism accepts that the universe has parts, but these move around more holistically in a continuous background medium, a plenum, that is either the aether or something more spiritual.

The aether has no voids. It is smooth and continuous. In Greek mythology, aether was the air breathed by the gods. Aristotle considered it to be the fifth element, after air, fire, water, and earth. He called it the *quintessence* and proposed that it was responsible for the motion of celestial bodies. In short, Aristotle's aether is not just another part of the universe; it is the central, governing part that unites everything into one harmonious whole.

Until the twentieth century, the atomists were the only natural philosophers who did not take it for granted that bodies in the universe swim about in an all-encompassing, continuous aether. For example, Descartes held that the planets were maintained in their orbits by swirling vortices in the aether.

In the 1680s, Newton rejected Descartes's vortex gravity in favor of a more mysterious action-at-a-distance force that he visualized in terms of the alchemic notion of an "active spirit." Alchemists did not distinguish matter from spirit and imagined God as guiding the process by which matter could be mutated from one form to another. Newton took gravity to be a similar action of the spirit of nature.[10]

So, quite ironically, Newton adopted the materialistic model of atoms and the void, doing away with the need for a continuous aether. But in order to explain gravity, he invoked the existence of another reality, one that was immaterial.

On the other hand, the wave theory of light was fully materialistic. It did not need any active spirit. But if light was an undulation, some medium had to be doing the undulating, and this was naturally taken to be the aether.

Newton had raised another objection to the aether model, pointing out that the planets would be slowed down and eventually brought to a halt by friction. So the ad hoc assumption had to be made that the aether, if it exists, must be frictionless.

In the nineteenth century, Michael Faraday (1791–1867) performed a host of intricate experiments demonstrating the properties of electricity and magnetism. These experiments led to important applications such as electric motors and generators and ultimately to a full understanding of these phenomena. One simple experiment of Faraday's is still widely used in elementary-school science demonstrations. Faraday sprinkled iron filings onto a sheet of paper held over variously shaped magnets, showing the "lines of force" that we associate with magnetic fields (see fig. 6.1). Faraday noted that gravity and electricity could be also visualized as acting by way of lines of force.

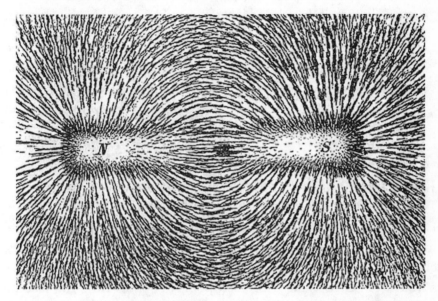

Figure 6.1. Iron filings sprinkled over a bar magnet. (Image from Newton Henry Black and Harvey N. Davis, *Practical Physics* [MacMillan, 1913], p. 343, fig. 200.)

Faraday proposed that the action-at-a-distance force that seems to occur when a body interacts by way of gravity, electricity, or magnetism actually takes place through intervening matter. However, it is important to note that Faraday did not view this intervening matter as a necessarily continuous fluid, as the aether is usually interpreted to be. He wrote:

> It appears to me possible, therefore, even probable, that magnetic action may be communicated by the action of intervening *particles* [emphasis added], in a manner having a relation to the way in which the inductive forces of static electricity are transferred to a distance (by transferring its action from one contiguous particle to the next).[11]

Faraday did associate the intervening matter with the aether (ether):

> Such an action may be a function of the ether; for it is not at all unlikely that, if there be an ether, it should have other uses than simply the conveyance of radiations.[12]

However, keep in mind that rather than a continuous fluid with no voids between them, as the Aristotelians claimed, Faraday's aether is actually composed of discrete particles, like the pebbles of sand on a beach.

FIELDS

Faraday and other nineteenth-century physicists applied a mathematical concept called a *field* to physical phenomena such as gravity, electricity, and magnetism. In physics, a field is a quantity that has a value, or set of values, for each point in space. Time can also be included, but let's put that off and just think of a field at a given time. The simplest field is a *scalar* field that has only one value at each spatial point. Examples of scalar fields are the pressure and density of a fluid, each of which require just one number to define it at each point.

The Newtonian gravitational field along with the classical electric and magnetic fields are examples of *vector* fields that have both a magnitude and a direction at each point in space. Since it takes two numbers to define a direction in three-dimensional space, a vector field requires three numbers to specify. In Einstein's general theory of relativity, the gravitational field is actually a *tensor* that needs sixteen numbers (not all independent) to define. But we need not get into that at this time.

Let's start with the Newtonian gravitational field at a point in space a distance r from a particle of mass m. By Newton's law of gravity, the field will have a magnitude that is directly proportional to m and inversely proportional to the square of r. The direction of the field will be along the line from the spatial point to the mass.

Now, we can see how to map out the field around the sphere. We just put a test particle at various points and watch the direction in which the particle accelerates. We can then sketch "lines of force" following these paths, as shown in figure 6.2. Note that we never actually see the field. All we see is the test particle accelerating through space.

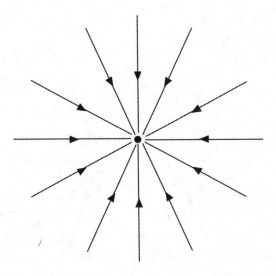

Figure 6.2. The lines of force representing the gravitational field surrounding a point mass. The lines show the path that a test particle will accelerate when placed at any point in space. The same figure represents the static electric field for a positively charged test particle when a negative point charge is at the center. If a positive charge is at the center, the lines point outward.

The same procedure can be used to map the electric field surrounding a point particle of electric charge q. Suppose q is negative. If we use a positively charged test particle, the field lines will again point from the test particle to q, as in figure 6.2. Here the magnitude of the field, given by Coulomb's law, is proportional to q and, as with the law of gravity, inversely proportional to the square of the distance r.

If the central charge is positive, the field lines will point outward. In either case, the field points in the direction in which a positive test particle accelerates.

Suppose instead of a single point charge we have two opposite charges. This is called a *dipole*. A single point charge is an *electric monopole*. We can use the same procedure of putting a test charge at various points. In this way, we map out the field shown in figure 6.3.

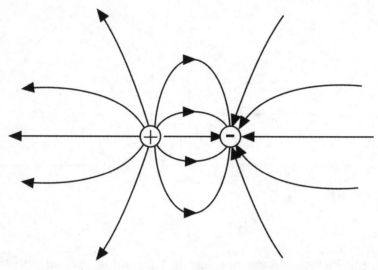

Figure 6.3. The lines of force representing the electric field surrounding an electric dipole, which is composed of two opposite point charges. They go from the positive to the negative charge.

The magnetic field is more difficult to define. The source of a magnetic field is an electric current—a moving electric charge. The magnetism of materials such as iron results from moving charges inside the material. However, Faraday showed how to map out the magnetic field. Magnetic filings are little bar magnets with North and South poles. A compass needle is a bar magnet. So you map a magnetic field by placing a small compass at various points and seeing the direction in which the needle points, as we see in figure

6.4. Note that because opposite magnetic poles attract, a compass needle will actually point to the *South Pole* of another magnet. That is, what we call the North Magnetic Pole of Earth is technically (at least in physics) the South Magnetic Pole!

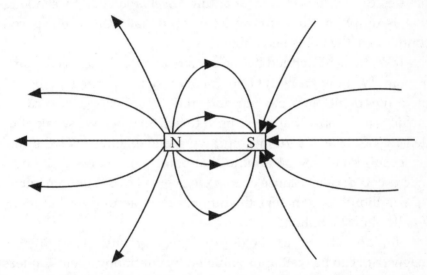

Figure 6.4. The magnetic field around a bar magnet. Note that it looks just like the electric dipole in figure 6.3. A bar magnet is a magnetic dipole.

We see that the magnetic field of a bar magnet looks just like that of the electric dipole in figure 6.3. That is, a bar magnet is a *magnetic dipole*. Then what happened to the *magnetic monopole*? No magnetic monopoles have ever been observed. If you have an electric dipole, such as a hydrogen atom, then you can pull the two charges apart and get two electric monopoles. However, if you cut a bar magnet in half, you get two bar magnets.

Modern particle theories predicted the existence of magnetic monopoles in the early universe. However, searches for them in the present have turned out negative. In chapter 12, we will see how this absence is now plausibly explained by inflationary cosmology.

The fields we have discussed—pressure, density, gravitational, electric, and magnetic—are each continuous in the assumed mathematical formalism. They are also regarded as functions of three coordinates needed to define a point in space—x, y, and z in the familiar Cartesian coordinate system. These coordinates are also assumed to be continuous variables, even in the most advanced physics discussions. Later, quantum mechanics would challenge that assumption. But for now, let us note that even when space is continuous, the fields need not be.

Take the field defined by the density of a fluid such as the air in a room. We now know that air is composed of particulate bodies, molecules of nitrogen, oxygen, and other ingredients such as carbon dioxide (more now than there used to be). So when we speak of the density of air at a certain point, we are averaging over the molecules in a small region around that point. This works well on the macroscopic scale because even if we average over a square micron (one-millionth of a meter), the number of molecules in that region will be almost a billion.

Similarly, the vector fields describing gravity, electricity, and magnetism can be well approximated by continuous fields, at least in everyday experience and in nineteenth-century laboratories.

ELECTROMAGNETIC WAVES

In 1865, James Clerk Maxwell proposed a set of equations that enabled the calculation of electric and magnetic fields for any arrangement of charges and currents. Maxwell's equations had a solution in a region absent of charges and currents that corresponded mathematically to that of a wave. Furthermore, the electromagnetic wave traveled precisely at the speed of light in a vacuum, which was a number that was not inserted independently into the theory but came out as a result.

Light was thus interpreted as an electromagnetic wave. Since no limits were placed on the wavelength of such a wave, the existence of waves of both lower and higher wavelengths than the visible spectrum was implied. In 1887, German physicist Heinrich Hertz (1857–1894) produced radio transmissions in the laboratory that traveled at the speed of light. Interpreted as waves, they had a wavelength of 8 meters, compared to that of visible light, which ranges from 400 to 700 nanometers, where a nanometer is one-billionth (10^{-9}) of a meter. Wavelengths in the electromagnetic spectrum range from one-hundredth of a nanometer and lower for gamma rays to a meter and above for radio waves. I have worked with telescopes that have detected gamma rays from space with apparent wavelengths on the order of one-billionth of a nanometer (10^{-18} meter). At the other end of the scale, observers at radio telescopes have detected radio signals that are interpreted as electromagnetic waves with wavelengths of many kilometers.

While the successful prediction of electromagnetic waves might be seen as the final confirmation of the existence of the aether, Maxwell claimed it as only a model. Unlike the theory of sound in which the mechanical vibration of a continuous medium is assumed as a starting point, no such assumption is made in Maxwell's theory. In fact, electromagnetic waves just fell out of the mathematics once he had placed the various independent principles of electricity and magnetism that had previously been discovered, such as Ampere's law, Gauss's law, and Faraday's law, into a single set of equations.

THE DEMISE OF THE AETHER

As I have already indicated, whether the aether exists is a major determining factor in the conflict between atomism and antiatomism. In one form or another, this issue has surfaced historically and, as we will see, still rises up on occasion even today.

For now, however, let us focus on the electromagnetic aether, the medium in which the electric and magnetic fields each are presumed to exist as some kind of tension analogous to that in the stretched skin of a drum. Electromagnetic waves, then, are viewed as analogous to the sound waves that would be set up when the drum is struck with a hammer.

If electromagnetic waves were analogous to sound waves, then the speed of light should vary depending on the relative motions of the source and observer through the aether. At the dawn of the twentieth century, two American physicists, Albert Michelson (1853–1931) and Edward Morley (1838–1923), attempted to detect the motion of Earth through the aether by comparing the speeds of light in two perpendicular beams. These speeds should have differed if light were a vibrating wave in the aether because the beams would be heading in different directions through the aether. Although Michelson and Morley were capable of measuring a speed difference of one-hundredth of that expected, they found no difference.

Now, it is important to note that while the Michelson and Morley result seemed to violate Galileo's principle of relativity, they in fact agreed with Maxwell's theory. Maxwell's equations do not say that the speed of light is $c + v$ when a source is moving toward you with a speed v, and $c - v$ when the source is moving away at that speed, as expected from our experience with sound waves (the Doppler effect). Maxwell's equations say that the speed of light in a vacuum is *always* $c = 299,792,458$ meters per second in any and all reference frames. (As we see below, this number is the speed of light *by definition*.) Michelson and Morley confirmed that this is indeed a fact. Evidently, electromagnetic waves travel in the atomist's void—a region completely empty of matter or substance of any kind.

But how can the speed of light be absolute? Doesn't the principle of relativity say that all velocities, and thus speeds, are relative? In 1905, Einstein had the flash of insight to ask what the consequences would be if both Galileo and Maxwell were correct; that is, the principle of relativity is valid, and the speed of light in a vacuum is absolute. The result was the special theory of relativity.

Recall the way I worded the principle of Galilean relativity. It does not say all velocities are relative. It says that you cannot distinguish between being at rest and being in motion at constant velocity.

Special relativity maintains that principle, but bodies moving near or at the speed of light behave very differently from those of everyday experience. The equations that describe "relativistic" motion are significantly modified from those of "nonrelativistic" Newtonian mechanics. That is not so say the latter are wrong. The relativistic equations all reduce to the nonrelativistic ones in the limit of speeds very much less than the speed of light.

In any case, the electromagnetic aether failed to be confirmed empirically, and it was shown to be not only unnecessary for understanding the phenomena of light and other electromagnetic waves but also inconsistent with both data and theory. Nevertheless, despite the failure to find evidence for the aether, the electromagnetic wave theory successfully described most of the observed behavior of light prior to the twentieth century. However, there were a few exceptions that, as we will see in the next chapter, led to quantum mechanics and a drastic revision of Newtonian mechanics.

TIME AND SPACE IN SPECIAL RELATIVITY

Einstein resolved the apparent contradiction between Galileo's principle of relativity and Maxwell's equations for electromagnetism by assuming they both were correct. In particular, the speed of light in a vacuum, c, is the same in all reference frames. But then, something had to go, and that was our commonsense notions of space, time, and matter.

Let us consider a clock in which a light pulse is sent back and forth between two mirrors, as shown in figure 6.5(a). Each time the pulse hits either mirror, we get one tick in the clock. Suppose the mirrors are 3 meters apart. Since $c = 0.3$ meter per nanosecond (1 nanosecond = one-billionth of a second), it then takes 10 nanoseconds to go between mirrors. Thus, each tick will occur 10 nanoseconds apart to an observer in a reference frame in which the clock is at rest. Let's take her reference frame to be that of Earth.

Now suppose that the clock is moving with a speed of 0.6 *c* with respect to the same observer on Earth. She will see the light pulse still moving at a speed *c*, but now it has to travel along a longer, diagonal path, and so takes longer to go between the mirrors. Applying the Pythagorean theorem to figure 6.5(b), it can be shown that the observer will hear the ticks on the moving clock every 12.5 nanoseconds. In that time, the clock has moved $(0.6\ c)(12.5) = 7.5$ nanoseconds.

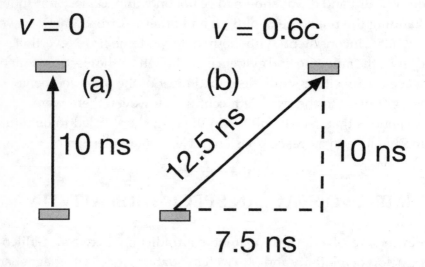

Figure 6.5. Example of time dilation. The time measured on a clock that is moving relative to an observer will give a greater time between two events and appears to slow down.

In other words, the time between two events separated in space depends on the reference frame of the observer. Time intervals are relative. In the above example, a clock that is moving with respect to an observer will measure a longer time between two events in that observer's reference frame. That is, it will appear to run slower. Its ticks will be farther apart than the clock at rest. This is called *time dilation*.

Since time intervals are relative, so must be space intervals. In the above example, the original observer on Earth measures the distance traveled by the clock in the time the pulse goes from one mirror to another to be its speed 0.6 c times 12.5 nanoseconds, which is 2.25 meters. However, to an observer in the clock's reference frame, the elapsed time is still 10 nanoseconds, since the clock is at rest in that reference frame. To him, Earth has traveled $(0.6\ c)(10) = 1.8$ meters.

In other words, an observer in one reference frame will see distances measured in a reference frame moving with respect to it contract in the direction of motion. Another way to look at it is to let a meter stick be attached to the ground. To the observer in the reference frame of the clock, the meter stick is moving at 0.6 c. It will appear shortened to $(1.8/2.25)(1) = 0.8$ meter. This is called *Fitzgerald-Lorentz contraction*.

Now, these phenomena are often mistakenly described as "a moving clock slows down," and "a moving object contracts," as if a moving clock physically slows down, or a moving object really contracts. But remember, motion is relative. Einstein maintained Galilean relativity, which says that all reference frames moving at constant velocity with respect to one another are equivalent. If an earthling sees a clock in a spaceship slow down, an observer in the spaceship will see no change in her clock. In fact, she will see the earthling's clock slow down.

When Einstein first presented this result, people thought it was paradoxical. How could both clocks run slower? This is called the *clock paradox* or, also, the *twin paradox*. If an astronaut goes off on a trip, she will be younger when she returns than her twin left back on Earth.

Einstein pointed out that the two observers can't compare their clocks without one turning around and coming back, that is, accelerating. Then the two frames are no longer equivalent.

In figure 6.6, I show how the clock paradox is resolved quantitatively. We have three clocks, one on Earth and one each on two spaceships, A and B. The vertical axis shows time measured on the Earth clock in years. The horizontal axis shows distances measured on Earth in light-years. (One light-year is the distance traveled by light in a vacuum in one year.)

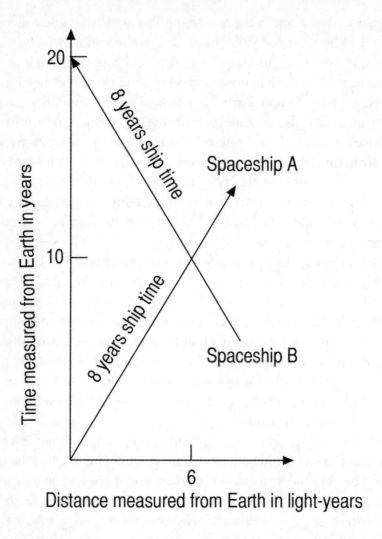

Figure 6.6. Resolution of the clock paradox. Spaceship A travels six light-years away from Earth in 8 years of ship time and 10 years of Earth time. Spaceship B returns at the same speed. So the total elapsed time is 16 years of ship time and 20 years of Earth time.

At time zero, Spaceship A moving at 0.6 c passes close to Earth. Its clock is synchronized with one on Earth. After 10 years, as measured on Earth, Spaceship A has traveled six light-years from Earth. However, the elapsed time measured on Spaceship A's clock is only 8 years.

At that instant, Spaceship B, heading toward Earth at the same speed of 0.6 *c*, passes close to Spaceship A and the clocks on the two ships are synchronized. Spaceship B reaches Earth 8 years later as measured on its clock, and so the total trip time for A's trip out and B's trip back is 16 years of ship time. However, 20 years have elapsed on Earth.

The difference is no more paradoxical than noting that a crooked path between two points has a different length than a straight path; except in time units, the straight path is the *longest* between two events.

DEFINING TIME AND SPACE

Notice that in all this discussion we never talk about what space and time "really" are. Most people will certainly assume they are aspects of reality, but we have only considered what is measured with clocks and meter sticks. That is, we have avoided getting into any metaphysical issues by just dealing with what we measure with our measuring devices. As we saw in chapter 2, Einstein said time is what you measure with a clock (although he may have changed his mind later). By international agreement, the basic unit of time is the *second*, which in 1967 was defined as 9,192,631,770 periods of the radiation corresponding to the transition between the two specific energy levels of the Cesium-133 atom.

Until 1983, the meter was defined separately from the second. By then, however, the special theory of relativity was so well established empirically that it was agreed to treat space and time on the same footing. So the meter was redefined as the distance between two points when the time it takes light to go between the points in a vacuum is 1/299,792,458 second. In other words, distance is no longer defined as what you measure with a meter stick. Like time, it is defined by a measurement on a clock.

Many people, including many scientists, still think the speed of light in a vacuum is something you must measure to obtain its value. Well, it is not. It is an arbitrary number that simply depends on what units you would like to use for length. In the metric system, the speed of light in a vacuum is $c = 299,792,458$ meters per second *by definition*. In the English system, $c = 983,571,056$ feet per second *by definition*.

Oh, you can go ahead and measure the speed of light in a vacuum if you want, with a meter stick and a clock. But you will have to get 299,792,458 meters per second because whatever device you use to measure distance will be calibrated to give 299,792,458 meters when whatever device you use to measure time measures exactly one second to the same precision.

If you are an astronomer, you might prefer to measure distance in light-years and then the speed of light in a vacuum is, by definition, one light-year per year. (Actually, many astronomers still use an antiquated unit called the *parsec*, which is 3.26 light-years.) Particle physicists also like to work in units where $c = 1$, which gets rid of a lot of useless c's in their equations.

By defining both distance and time in terms of measurements made with a clock, space and time were placed on the same footing. Time is now treated as another dimension added to the three dimensions of space. Rather than call it the fourth dimension, we count from zero so that time is the "zeroth" dimension. So we can think of the position of an event occurring at coordinates x, y, and z in three-dimensional space and at time t in the zeroth dimension of time, combined as a *4-vector* (t, x, y, z).

MATTER AND ENERGY IN SPECIAL RELATIVITY

Einstein showed that the relativity of time and distance requires substantial changes in the way we describe matter and motion. I will not attempt to prove the equations that replace those from nonrelativistic physics ($v \ll c$), but will just give the results. The three quantities that define the properties of a material body and its motion are mass, energy, and momentum. Mass is the measure of a body's *inertia*, that is, its sluggishness or resistance to changes in its motion. The more massive a body is, the greater the force that must be applied to accelerate it to a higher speed or decelerate to a lower speed. When a body is moving, its internal clocks, such as atomic motions, will be observed to slow down. Thus, such a body will appear to be more sluggish and thus more massive. That is, a moving body will have a greater measured mass than one at rest, as measured in the reference frame in which it is moving.

Einstein associated the mass of a body at rest with a certain quantity called *rest energy,* that is, $E_0 = mc^2$, which popular writers always refer to as "Einstein's famous equation."[13] Now, we saw that c is just an arbitrary constant, so let us simply take $c = 1$. Then the mass of a body, m, is the same as its rest energy.

So a body at rest has an energy m. When it is moving, the body has an additional energy K we call the *kinetic energy.* Then the total energy of a free body, that is, one without any forces acting on it, is $E = m + K$.

The quantity of motion that was used by Newton in his laws of motion is the momentum p (just the magnitude is needed here). The three quantities of matter and motion are related by $m^2 = E^2 - p^2$. That is, only two are independent variables. The mass m is invariant, that is, it is the same in all reference frames. E and p depend on frame of reference. (Recall discussion of frames of reference in chapter 3.) However, in any given frame, both are conserved in the absence of any external forces, as they are in Newtonian physics.

Let the *3-momentum* of a particle be defined as (p_x, p_y, p_z), where p_x is the component of the momentum projected on the x-axis, and so on. Just as in the previous section, where we defined the 4-vector position of an event in space-time, we can define the *4-momentum* of a particle as (E, p_x, p_y, p_z), where E is the particle's energy. That is, like time and space, energy and momentum are on the same footing. Matter is then defined as anything with a 4-momentum.

In the following chapters on atomic, nuclear, and particle physics, I will use units in which $c = 1$. In that case, mass, energy, and momentum all have the same units. The basic unit will be the electron volt (eV) and multiples of it (keV, MeV, GeV, TeV). This is a unit defined as the kinetic energy an electron gains when it falls through an electric potential of 1 volt. The energies involved in atomic physics and condensed matter physics and chemistry are typically in the eV and keV range. Nuclear energies are typically in the MeV range (1 MeV = 10^6 eV). Particle-physics energies are usually measured in GeV (1 GeV = 10^9 eV) or, more recently, TeV (1 TeV = 10^{12} eV). In these "energy units," the mass of an electron is 0.511 MeV and the proton is 938 MeV.

INVARIANCE

When a quantity has the same value in all reference frames, it is said to be *invariant*. In this chapter, we have seen that the speed of light in a vacuum and the mass of a body are invariant. On the other hand, space and time intervals, energy, and momentum are not invariants.

In physics, we express our mathematical models in terms of measurable quantities, such as distance, time, energy, and momentum. If we want those models to be objective and universal, they should not change from reference frame to reference frame. That is, we would like our models and theories to be invariant.

Consider a simple example. In a given reference frame, the distance x a photon will travel in a vacuum in a time t will be $x = ct$. In another reference frame moving at a constant speed with respect to the first, the distance x' a photon will travel in a vacuum in a time t' will be $x' = ct'$. Notice that the distance and times are different, which is why I put primes on their symbols. But since this is a photon in a vacuum, its speed is the same in each reference frame, that is, c with no prime. Thus, we can write $x - ct = x' - ct' = 0$. That is, the equation $x - ct = 0$ is an invariant model describing photon motion in all reference frames. The notion of invariance played a key role in the development of twentieth-century physics.

SYMMETRY

Let us do a flashback at this point and discuss the ever-increasing recognition that symmetry principles are the foundation of physics. Recall that symmetry and invariance are related concepts. When an object looks the same from all angles, we say it is rotationally invariant and possesses spherical symmetry. When it looks the same in a mirror, we say it is reflection invariant and possesses mirror symmetry.

Similar statements can be made about the mathematical models of physics. Consider a model that describes observations in terms of space and time. (Most do, but not all). The spatial positions of particles, for example, might be described by a set of Cartesian coordinates (x, y, z) at various times t. If the equations in the model do not change when you move the origin of the coordinate system from one place to another, then the model is *space-translation invariant* and possesses *space-translation symmetry*. If the equations in the model do not change when you rotate the axes of the coordinate system, then the model is *space-rotation invariant* and possesses *space-rotation symmetry*. If the equations in the model do not change when you change the time you start your clock, then the model is *time-translation invariant* and possesses *time-translation symmetry*.

Recall the principle of Galilean relativity. We can regard that as a symmetry principle. It implies that our physics models must be the same in all reference frames—invariant—moving at constant velocity with respect to one another.

When we move from Galileo to Newton we find that Newton's laws of motion and gravity possess all the symmetries mentioned so far, including mirror symmetry. The laws of thermodynamics, which, as we have seen, follow from Newtonian mechanics, also possess these symmetries. The same holds for the laws of electrodynamics, which also are invariant to a change in the sign of electric charge—what is called *charge-conjugation symmetry.*

When we get to the twentieth century, we have special relativity in which Galilean invariance is replaced by Lorentz invariance, a different equation but the principle is the same. Einstein introduced the term *general covariance* to describe the principle in which physical models are invariant to arbitrary coordinate transformations of the type we have been discussing. In general relativity, he extended that to include accelerated reference frames.

THE SOURCE OF CONSERVATION PRINCIPLES

One of the most import contributions to twentieth-century physics was made by a German mathematician named Emmy Noether, who is only now getting the full recognition that she deserves.[14] In 1915, Noether proved the following theorem.[15]

Noether's Theorem:

> For every continuous symmetry of the laws of physics, there must exist a conservation law.
>
> For every conservation law, there must exist a continuous symmetry.

This is not only a profound physics result; it is a profound philosophical one. It means that the most important of the so-called laws of physics are not what is commonly believed they are, that is, that they are rules built into the structure of the natural world that govern the behavior of matter. Rather, each is an automatic ingredient of any physical theory that possesses the associated symmetry.

Here are the quantities that are conserved with each specific symmetry:

Time-translation symmetry	conservation of energy
Space-translation symmetry	conservation of linear momentum
Space-rotation symmetry	conservation of angular momentum

For example, if a physicist builds a mathematical model that possesses time-translation symmetry, then that model will necessarily obey conservation of energy.

Noether's theorem can be looked at another way. If a physical system does not conserve one of these quantities, then that system will not possess the corresponding symmetry. For example, a body dropped from the top of a tower gains linear momentum at a fixed rate; that is, linear momentum is not conserved. A series of video frames taken at various intervals along its path to the ground would not be identical. The first frame might show the body at rest. Each succeeding frame will show the body moving faster than the frame before.

So the conservation principles will not necessarily hold for every physical system. However, over the centuries, they have been observed to hold for isolated systems, that is, those that do not interact in some way with other systems. If we consider Earth and the body dropped from the tower, then in that system, momentum is conserved.

When we consider the universe as a whole, then all known conservation principles seem to hold. We conclude from this that our universe comprises an isolated system and possesses the associated symmetries. This means there is no special moment in time (the universe could have appeared at any time), no special place in space (the universe can be anyplace), and it is rotationally symmetric (the universe looks the same from all directions) so no special direction in space is singled out.

Thus, the validity of the three great conservation laws of physics is testimony to a universe that is isolated from anything on the outside and looks just like it should look if it came from nothing.

7

INSIDE THE ATOM

There is no quantum world. There is only an abstract physical description. It is wrong to think that the task of physics is to find out how nature is. Physics concerns what we can say about nature ...
—**Niels Bohr**[1]

ANOMALIES

In the late nineteenth century, the wave theory of light seemed secure. It was consistent with all observations with just three exceptions, three anomalies that could not be explained within the theory as it was understood at the time.

When the optical spectrograph was developed, it became possible to measure the wavelengths of light from various sources. These measurements extended beyond both ends of the visible region of the electromagnetic spectrum, from the shorter wavelengths of the ultraviolet to the longer wavelengths of the infrared.

It was found that all bodies emit a smooth spectrum of light, whose peak intensity occurs at a wavelength that depends on the temperature of the body: the higher the temperature, the shorter the wavelength at peak intensity. For the very hot sun, the peak is smack in the center of the visible spectrum, in the region humans recognize as the color yellow.

What a coincidence! The spectrum of our main source of light, the sun, just happens to be right in the center of our visual capability. This is an example of what in recent years has been termed the *anthropic principle*: the universe is just so, "fined-tuned" so we and other life-forms are able to exist.

Actually, the universe is not fine-tuned for us; we are fine-tuned to the universe. Obviously, our eyes evolved so they would be sensitive to the light from the sun. Animals that could see only x-rays would find it difficult to survive on a planet where most of the objects around them do not emit x-rays. While other so-called anthropic coincidences are not so obvious, it has not been shown that they are so unlikely that the only explanation is a supernatural creation.[2]

At lower temperatures, such as those in our everyday experience, bodies radiate light in the infrared region where our eyes are insensitive, although we can see the infrared with the night-vision goggles of modern military technology or with infrared photography. You can buy a photo of your infrared "aura" at most psychic fairs. It is not empirical evidence for any supernatural qualities.

Despite psychic claims, your aura is perfectly natural. It is an example of what is called *blackbody radiation*. Everyday objects that do not emit or reflect visible light appear black to the naked eye, hence the name of this type of radiation.

According to the wave theory of light, vibrating electric charges inside a body emit electromagnetic waves. The lower the wavelength of these waves, the greater number of them can "fit" inside the body. So the wave theory predicts that the intensity of emitted light will increase indefinitely as you go to shorter and shorter wavelengths. This is called the *ultraviolet catastrophe* and is simply not observed. Rather the intensity of light falls off smoothly at both ends of the spectrum. This is anomaly number one.

Anomaly number two also involves the spectra of light. When a gas is subjected to an electric discharge (a spark) or otherwise heated to a high temperature, the observed spectrum is not totally smooth, although it still has a smooth part, but is composed of very narrow, bright lines at well-defined wavelengths. These are called *emission lines*. A gas will also exhibit sharp, dark *absorption lines* when a broad spectrum of light shines through it. The wonder of these line spectra is that they are different for gases of different chemical composition, providing a powerful tool for determining chemical compositions for materials on Earth and even deep in space.

The third anomaly that was observed in the late nineteenth century is the *photoelectric effect*. When light is beamed onto the metal cathode of a vacuum tube, an electric current is sometimes produced in the tube. The wave theory predicts that the current should increase with increased intensity of light, since the energy in a light beam is greater when the intensity is greater. In fact, increasing the intensity has hardly any effect. Instead, a threshold frequency exists below which no current is produced, a current occurring only after you go above that value. In addition, the energy of the free electron is a function of the frequency of the light, not its intensity.

The frequency f of a wave is the number of crests of the wave that pass a certain point per unit time. The wavelength λ of the wave is the distance between crests. The two are related by $f = c/\lambda$, where c is the speed of the wave. In the case of light in a vacuum, c is the speed of light.

LIGHT IS PARTICLES

At the University of Munich, Max Planck (1858–1947) had written his doctoral dissertation, which he defended in 1879, on the second law of thermodynamics. He was no immediate fan of atomic theory, having the same objection as many physicists of the time that it involved probabilities rather than the certainties of thermodynamic law. In 1882 he wrote, "The second law of the mechanical theory of heat is incompatible with the assumption of finite atoms. . . . A variety of present signs seems to me to indicate that atomic theory, despite its great successes, will ultimately have to be abandoned."[3]

Sometime around 1898, Planck seems to have made an abrupt reversal in his thinking and realized that the second law must be probabilistic. He had been studying the thermodynamics of electromagnetic radiation and realized that the same probability arguments used to derive Boltzmann's formula for entropy, $S = k\log W$, could be applied in that situation.[4]

In 1900, Planck proposed a model that quantitatively described the blackbody spectrum. He still applied the wave theory in visualizing a body as containing many submicroscopic oscillators. However, he added an auxiliary assumption: the energy of the electromagnetic waves occurs only in discrete bunches proportional to the frequency of the waves. Planck dubbed these bunches *quanta*. While he was not quite ready to call these "atoms of light," he had made a major move in applying the discreteness of the atomic theory to electromagnetism and thereby ushered in the quantum revolution.

Planck proposed that the energy of quanta are multiples of hf, where f is the frequency and h is a constant of proportionality now called *Planck's constant*. He was able to estimate the value of h by fitting his model to measured blackbody spectra. Not only did he get the shape of the spectrum right, he also obtained the correct temperature dependence. In the energy units we are using, $h = 4.14 \times 10^{-15}$ eV-seconds. That is, the energy of a quantum of light with a frequency $f = 10^{15}$ cycles per second (Hertz), or a wavelength of $\lambda = c/f = 3 \times 10^{-7}$ meter, will be 4.14 eV.

Planck did not speculate on the origin of the quantization of light energy. Einstein provided the explanation in a remarkable paper on the photoelectric effect that was published the same year, 1905, as his other remarkable papers on special relativity and Brownian motion. Although you will hear otherwise from some authors who would like you to think that the quantum revolution dispensed with the reductionist, materialistic view of matter and energy, we will see that the story is quite the opposite.

It was Einstein who proposed that the quanta of light were actually particles, later dubbed *photons*. Each photon has an energy

$E = hf$, where f is the frequency of the corresponding electromagnetic wave, as suggested by Planck. Einstein viewed the photoelectric effect as a particle-particle collision between a photon and an electron in the metal (see fig. 7.1). Since that electron is initially bound in the metal, some minimum energy is needed to kick it out into the vacuum tube to produce the observed current.

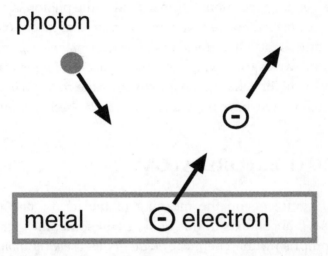

Figure 7.1. Einstein's explanation of the photoelectric effect. A photon with sufficient energy collides with an electron bound inside a metal and kicks it out of the metal. If the metal is the cathode of a vacuum tube, an electric current is produced.

An American physicist, Robert A. Millikan (1868–1953), was convinced that Einstein was wrong since it was already well established that light was a wave. He spent a decade trying to prove it. Instead, as he later admitted, his results scarcely permitted any other explanation than the photon theory of light.

By applying a back potential to the anode of a vacuum tube, Millikan was able to measure the minimum energy a photon had to have in order to remove an electron from the metallic cathode, as a function of frequency of the corresponding light wave. He found that the energy was proportional to the frequency of the light with a constant or proportionality equal to Planck's constant, precisely as Einstein had predicted.

The equations of special relativity require that an object traveling at the speed of light must have zero mass. However, a photon has both energy and momentum, which (in units $c = 1$) implies that $m^2 = E^2 - p^2 = 0$, that is, $E = p$. So, even though it has zero mass, since anything with energy and momentum is matter, the photon is matter.

And so, Einstein revived Newton's corpuscular theory of light. At the same time, light still diffracted as expected from waves, and no violation of Maxwell's electromagnetic wave theory of light was found. Light thus exhibits what became known as the *wave-particle duality*, about which we will have more to say. But first—back to atoms.

THE RUTHERFORD ATOM

To avoid confusion on the meaning of the term *atom*, which is used in several contexts in this book, I will sometimes refer to the atoms comprising the chemical elements as *chemical atoms*. Early in the twentieth century, the chemical atoms were found not to be uncuttable after all but composed of parts that are more elementary.

In 1896, French physicist Henri Becquerel (1852–1908) discovered that uranium wrapped in thick paper blackened a photographic plate. The paper ruled out the possibility that the effect was caused by visible light, such as from phosphorescence. The phenomenon was dubbed *radioactivity*. Further observations, notably by Marie (1867–1934) and Pierre (1859–1906) Curie, identified various types of radioactive emanations associated with different materials. These were called alpha, beta, and gamma radiation. Ernest Rutherford (1871–1937) inferred that some of these phenomena were associated with the transmutation of one element to another.

In 1909, Rutherford, along with Hans Geiger (1882–1945) and Ernest Marsden (1889–1970), performed an experiment in which gold foil was bombarded with alpha radiation. In 1911, Rutherford proposed a model of the gold atom that explained the large angles at which the alpha rays were occasionally deflected. In Rutherford's model, the atom has at its center a positively charged nucleus that contains almost all the mass of the atom while at the same time being very much smaller than the atom as a whole. Rutherford envisaged the atom as a kind of solar system, with the nucleus acting as the sun and electrons revolving around the nucleus like planets.

In 1917, Rutherford made the alchemists' dream a reality. He did not make gold, but instead succeeded in transmuting nitrogen into oxygen by bombarding nitrogen with alpha rays. He determined that the nucleus of hydrogen was an elementary particle, which he called the *proton*. Rutherford speculated that the nucleus also contains a neutral particle, the *neutron*, which was eventually confirmed in 1932 by his associate James Chadwick (1891–1974). Eventually the three types of nuclear radiation were identified as particulate: alpha rays are helium nuclei (a highly stable state of two protons and two neutrons); beta rays are electrons; gamma rays are very high-energy photons.

With the discovery of the neutron, all of matter could be described in terms of just four elementary particles. All the chemical elements and their compounds, which formed all the material substances known at that time, were composed of a tiny nucleus, made of protons and neutrons, surrounded by electrons. However, as we will see next, these electrons are no longer viewed as orbiting "planets" but rather as a diffuse cloud.

The fourth elementary particle known at that time was the photon. As already noted, the photon has zero mass but still carries energy and momentum. In 1916, Einstein showed in his general theory of relativity that the photon is acted on by gravity, another proof that it is material in nature. You just have to be careful to use the modified kinematical equations of special relativity when you are describing a photon's motion, since it travels at the speed of light.

THE BOHR ATOM AND THE RISE
OF QUANTUM MECHANICS

In 1913, Niels Bohr (1885–1962) used Rutherford's atomic model to develop a quantitative theory of the simplest atom of all, hydrogen, which is composed of an electron in orbit about a proton. For simplicity, Bohr initially assumed the electron followed a circular orbit. However, the planetary model of the atom had a serious problem. According to classical electromagnetic theory, an accelerated charged particle will radiate electromagnetic waves and thereby lose energy. Since orbiting electrons continually change direction, they are accelerating and so will quickly spiral into the nucleus, collapsing the atom.

Bohr's solution was to hypothesize that only certain orbits are possible and that each orbit corresponds to a different, discrete "energy level." In the lowest orbit, the electron has a minimum energy and cannot go any lower, thus making the atom stable. When the electron is in a higher orbit, it can drop down to a lower one, emitting a photon with energy equal to the energy difference between levels. In this way, only certain well-defined energies are emitted, giving the line spectrum of wavelengths that are observed.

To specify which orbits are allowed, Bohr made the following hypothesis: the angular momentum of the electron must be in integral multiples of $h/2\pi$.[5] In physics, this quantity is called the *quantum of action* and is given the special symbol $\hbar$, called "h-bar." From this, Bohr was able to calculate the spectrum of hydrogen observed in spectrographic measurements. The theory was also applied to heavier atoms, called *ions*, in which all but one electron has been removed.

The energy levels of the hydrogen atom calculated by Bohr are

$$E_n = \frac{-13.6}{n^2} \text{ eV},$$

where n is the *orbital quantum number*, with integer values 1, 2, 3, and so on. Note that E_n does not include the rest mass of the electron, as is the convention in nonrelativistic physics. The negative sign means that this much energy ($|E_n|$) must be provided to remove the electron from the atom. When an electron drops from a higher energy level to a lower one, the emitted photon has an energy equal to the energy difference between levels. This produces the emission spectrum. The absorption spectrum results when photons with an energy equal to the difference between two levels are absorbed by lifting electrons from lower to higher levels.

However, the Bohr model could not explain the *Zeeman effect*, which is the splitting of spectral lines that occurs when atoms are placed in a strong magnetic field. Bohr and Arnold Sommerfeld made the obvious improvement, allowing for elliptical orbits analogous to Kepler's planetary system. They also took into account special relativity. The new model was characterized by three quantum numbers:

$n = 1, 2, 3, \ldots$ orbital quantum number

$\ell = 0, 1, 2, 3, n - 1$ azimuthal quantum number

$m = -\ell, -\ell + 1, -\ell + 2, \ldots, \ell - 2, \ell - 1, \ell$ magnetic quantum number

The orbital quantum number n is the same as in the Bohr model and in the absence of external fields; the energy levels are the same, depending only on n. The azimuthal quantum number ℓ is a measure of the ellipticity or shape of the orbit; the smaller the value, the more elliptical. The $\ell = 0$ orbit goes in a straight line right through the nucleus and back. The magnetic quantum number m is a measure of the orientation or tilt of an orbit.

As a circulating charge, the electron produces a magnetic field that will be different for different orbital shapes and tilts. The spectral line shifts observed in the presence of external magnetic fields result from the interaction energies between the external field and the fields produced by circulating electrons in orbits with different tilts.

The Bohr-Sommerfeld model successfully described the Zeeman effect, but as experiments became increasingly refined, a "fine structure" of spectral lines was observed even in the absence of external fields that the model could not explain. In order to fit that data, the old quantum theory of Bohr and Sommerfeld and the planetary model was discarded and a new quantum theory was born.

ARE ELECTRONS WAVES?

Louis-Victor-Pierre-Raymond, 7th duc de Broglie (1892–1987), made the first step toward the new quantum theory in 1924. Although light had been shown to be composed of particulate photons, a beam of light still exhibited the diffraction and interference effects associated with waves. De Broglie noted that the wavelength associated with a photon with a momentum p is given by $\lambda = h/p$. He then conjectured that the same relation holds for electrons and other particles. That is, an electron of momentum p will have a "de Broglie wavelength," $\lambda = h/p$.

In 1927, American physicists Clinton Davisson (1881–1958) and Lester Germer (1896–1971) ostensibly verified de Broglie's conjecture when they measured the diffraction of electrons scattering from the surface of a nickel crystal. Since then, diffraction has been observed for other particles such as neutrons and even large objects such as buckyballs, molecules containing sixty carbon atoms.[6] In fact, every object—even you and I—has an associated wavelength. It's just too tiny to be observable for macroscopic objects because their momenta are so large.[7]

The wave-particle duality thus seems to hold for all particles, not just photons. However, as we will see, it is a mistake to think that an object is "sometimes a particle and sometimes a wave." The wave nature that is associated with particles applies to their statistical behavior and should not necessarily be assumed to apply to individual particles. Neither photons nor electrons individually exhibit wavelike properties. Only beams of many of them do.

THE NEW QUANTUM MECHANICS

In 1926, Erwin Schrödinger (1887–1961) formulated a mathematical theory of quantum mechanics based on the classical mechanics of waves. He derived an equation that enables one to calculate a quantity ψ called the *wave function* for a particle of mass m, total energy E, and potential energy V. The wave function is a field with a value at every point in space and time. When E is positive, ψ is sinusoidal with a wavelength given by the de Broglie relation. However, it is important to remember that the wave function is not always in the form of a wave.

E can also be negative when the particle is bound, such as an electron in the hydrogen atom. Schrödinger was able to solve his equation for the hydrogen atom and obtain the same energy levels derived by Bohr.

The wave function in the Schrödinger equation is actually a complex number, so it has two values at each point in space-time—an amplitude A and a phase ϕ. The equation itself is a partial differential equation utilizing mathematics that is taught at the undergraduate level and is familiar to any natural science, engineering, or mathematics major.

However, Schrödinger was not the first to develop the new quantum mechanics. In the previous year, 1925, Werner Heisenberg (1901– 1976) had proposed a formulation using less familiar mathematical methods than those used by Schrödinger. A few months later in the same year, Heisenberg, Max Born (1882–1970), and Pascual Jordan (1902–1980) elaborated on Heisenberg's original ideas utilizing matrix algebra, which is not terribly more advanced than partial differential equations. In 1926, Wolfgang Pauli (1900–

1958) showed that matrix mechanics also yielded the energy levels of hydrogen.

In 1930, the two theories—Schrödinger wave mechanics and Heisenberg-Born-Jordan matrix mechanics—were shown by Paul Dirac (1902–1984) to be equivalent. Dirac's own version of quantum mechanics was based on linear vector algebra, which, while less familiar than Schrödinger's, is by far the most elegant and more widely applicable.[8] In the Dirac theory, the quantum state of a system is called the *state vector*, and the wave function is just one specific way to mathematically represent the state vector. Neither Heisenberg's or Dirac's quantum mechanics is based on any assumptions about waves.

Let me give some more details on the nature of the solutions of the Schrödinger equation for the hydrogen atom. The mathematical form of the wave function depends on three quantum numbers:

$n = 1, 2, 3, \ldots$	principle quantum number
$\ell = 0, 1, 2, 3, n - 1$	orbital angular momentum quantum number
$m = -\ell, -\ell + 1, -\ell + 2, \ldots, \ell - 2, \ell - 1, \ell$	magnetic quantum number

Note that the Bohr-Sommerfeld azimuthal quantum number has been renamed the orbital angular momentum quantum number. In the new quantum mechanics, the angular momentum L of an orbiting electron is quantized according to $L = \sqrt{\ell(\ell+1)}\hbar$ while the magnetic quantum number gives the component of angular momentum along any particular direction you choose to measure it. Let's arbitrarily call that the z-axis. Then $L_z = m\hbar$.

SPIN

In 1925, Pauli made a further improvement to the atomic model when he proposed that electrons and other particles have an intrinsic angular momentum, called *spin*. He added a new *spin quantum number s*, where the spin S of a particle is given by $S = \sqrt{s(s+1)}\hbar$ and its component along the z-axis is $S_z = m\hbar$. As with the magnetic quantum number m, m_s ranges from $-s$ to $+s$ in unit steps. In the case of the electron $s = \frac{1}{2}$ and $m_s = -\frac{1}{2}$ or $+\frac{1}{2}$.

Particles with half-integer spins are called *fermions*; those with zero or integer spins are called *bosons*. Pauli introduced what is now called the *Pauli exclusion principle*: only one fermion at a time can exist in a given quantum state. As we will see, this helped explain the periodic table of the chemical elements. The Pauli principle does not apply to bosons. Indeed, they tend to congregate in the same state, producing interesting quantum effects such as boson condensation.

At first, it was imagined that the electron was a solid sphere spinning about an axis. A spinning, charged sphere has a magnetic field caused by the circulating currents in the sphere. Furthermore, as we have seen in the Bohr-Sommerfeld model, the electron's motion around the nucleus also generates a magnetic field, so the two fields can interact with one another, producing a split in the spectral lines of an atom even in the absence of an external field. However, it turned out that the spinning sphere model of the electron gives a magnetic field for an electron that is too small by a factor of two.

With the discovery of electron spin, quantum mechanics moved into a new realm. Up to this point, quantum phenomena all had classical analogues, such as the planetary atom. Spin was a whole different animal with no classical analogue. Points don't spin. Quantum mechanics cannot be fully understood with images solely based on familiar experience. That does not mean, however, as is often thought, that quantum mechanics cannot be understood.

DIRAC'S THEORY OF THE ELECTRON

In 1928, Paul Dirac developed a theory of the electron that took into account special relativity. Remarkably, the spin of the electron did not have to be inserted manually, as with Schrödinger's and Heisenberg's nonrelativistic theories, but fell right out of the mathematics. So did the factor of two in the electron's magnetic field, as measured by what is called the *magnetic dipole moment* of the field, which was missing from the model of an electron as a spinning sphere.

With the Dirac equation for the electron, the detailed spectrum of hydrogen, as it was measured at the time, was fully described. However, just after World War II, in the aftermath of the Manhattan Project, incredibly precise experiments led to a theoretical break-through called *quantum electrodynamics* (QED). Quantum electro-dynamics integrated Dirac's relativistic quantum mechanics into a new framework known as *relativistic quantum field theory*, which will be discussed in chapter 9.

Dirac's theory was even more world shaking than all that. It had negative energy solutions that he associated with antiparticles, in particular, antielectrons. These are partners of electrons that are identical in every way except their electric charges are opposite. In 1932, Carl D. Anderson (1905–1991) observed positive electrons in cosmic rays and dubbed them *positrons*. Earlier the same year, Chadwick had discovered the neutron, which, we recall, suggested that the physical world was composed of just four particles: the photon, electron, proton, and neutron. Now, a few months later, we had to add the positron. Eventually physicists would find antipro-tons and antineutrons. The universe is simple, but not too simple.

The Dirac equation applies only to spin ½ fermions. However, other relativistic quantum equations were shortly developed for spin 0 and spin 1 bosons. These are all that are necessary for most purposes because all known elementary particles have either spin ½ or spin 1. As we will see, a spin 0 particle called the *Higgs boson* that is part of the so-called standard model of elementary particles and forces has now apparently been observed.

WHAT IS THE WAVE FUNCTION?

Relativistic quantum field theory is needed to understand the structure of matter at its deepest levels. The wave function that is prominent in the Schrödinger theory plays no important role in quantum field theory or in relativistic quantum mechanics. It is mentioned on only one page in Dirac's classic 1930 book, in a dismissive footnote:

> The reason for this name [wave function] is that in the early days of quantum mechanics all the examples of these functions were in the form of waves. The name is not a descriptive one from the point of view of the modern general theory.[9]

However, Schrödinger wave mechanics remains in common use and is by far the most familiar to physicists, chemists, biologists, and others who work on lower-energy, nonrelativistic phenomena such as condensed matter physics, atomic and molecular chemistry, and microbiology. So, let us ask: What is this wave function? What does it mean?

No one has ever measured a wave function. Schrödinger thought that A^2, the square of the amplitude of the wave function, gives the charge density of a system such as an atom. De Broglie proposed it was a "pilot wave" that guided particles in their motion.[10] This view was developed further by David Bohm in the 1950s.[11]

But the interpretation of the wave function that caught on and is still accepted by the consensus was the one proposed by Max Born in 1926. Born hypothesized that the square of the amplitude of the wave function at a given point in space and time is the probability per unit volume, that is, the probability density, for observing the particle at that position and time. In this picture, the wave function is not some kind of waving field of matter any more than the electromagnetic field is a waving of the aether. It is a purely abstract, mathematical object used for making probability calculations. Furthermore, we no longer picture the atom as a submicroscopic solar system, but as a nucleus surround by a fuzzy cloud of electrons. Wherever the cloud is denser, the more likely an electron will be found at that point.

In this picture, the de Broglie wavelength is not, as usually described, the "wavelength of a particle." More precisely, it is the wavelength $\lambda = h/p$ of the wave function, the mathematical quantity that allows you to calculate the probability of a particle of momentum p being detected at a certain position at a certain time. Put another way, if you have a beam of many particles, each with a momentum p, it will behave statistically like a wave of wavelength $\lambda = h/p$, for example, by exhibiting diffraction, interference, and other wavelike behavior.

THE HEISENBERG UNCERTAINTY PRINCIPLE

In 1927, Heisenberg asked what would happen if you tried to measure with increasing accuracy the position of a particle. To do so, you would have to scatter light off the object with shorter and shorter wavelengths, or else diffraction effects would wipe out any position information.

Assume, for simplicity, that the light beam is monochromatic, that is, it has a well-defined wavelength λ. Heisenberg noted that, according to de Broglie, the light beam would contain photons of momenta $p = h/\lambda$. The shorter the wavelength, the higher will be the corresponding momentum.

When the photons scatter off the object in question, a good portion of their momenta will be imparted to the object, rendering its momentum increasingly uncertain. Heisenberg determined that the product of uncertainty in a particle's position Δx and the uncertainty (technically, the standard error) in its momentum Δp can never be zero. The product must always be greater than or equal to $\hbar / 2$.

$$\Delta p \, \Delta x \ge \frac{\hbar}{2}$$

In the case of an electron in an atom, if you tried to measure its position accurately within the atom, you would have to hit it with photons or other particles of such high momentum that they would likely kick the electron out of the atom. Hence, the electrons in an atom are best described as a cloud in which their positions are indefinite.

Now, you might say that the electrons still move around in definite orbits and we just can't measure them. This is an example of the new philosophical questions that arise in quantum mechanics. If you can't measure a particle's position accurately, does it "really" have a well-defined position? The answer seems to be no.

Take, for example, the way atoms stick together to form a molecule. Consider the simplest case, the hydrogen molecule composed of two hydrogen atoms. You can think of one atom being a positive H^+ ion, which is simply a proton with the electron removed. This positive ion comes close to a negative H^- ion, a hydrogen atom with an extra electron. The two will attract each other by the familiar, classical electrostatic force.

However, there is more to it than that. Because of the uncertainty principle, the electron's position cannot be measured with sufficient accuracy to determine which atom it is in. Thus, it is in neither—or both. This results in an additional attractive force called the *exchange force* that, like spin, has no classical analogue.

Now, if you persist in trying to understand quantum mechanics in terms of classical images, or worse, try to relate it to some ultimate "reality," you have some work to do. How can the electrons be in both atoms at once? Once again, if we try to measure their position with, say, a photon of sufficiently high momentum, we will split the molecule apart. Is it proper to even talk about something you cannot measure? Just because you can't measure something, does that mean it can't be real? Why can't it really be real and we just can't measure it? The best answer we can provide is the one adopted by most physicists today: "Shut up and calculate." We have a model that agrees with observations and enables us to make predictions about future observations. What else is needed?

Now, the above development of the uncertainty principle followed the traditional derivation as originally proposed by Heisenberg. However, the uncertainty principle can be shown to follow from the basic axioms of quantum mechanics.[12]

BUILDING THE ELEMENTS

The chemical elements were the atoms of the nineteenth century. Recall that I refer to them as *chemical atoms*, since they are now known not to be elementary but composed of nuclei and electrons. The number of protons in a nucleus of a given chemical atom is given by Z, where Z is the atomic number that specifies its position in the periodic table (see fig. 4.1). That is, each chemical element is specified by the number of protons in the nucleus of the corresponding chemical atom. An electrically neutral atom has Z electrons to balance the charge of the nucleus. If it has more than Z electrons, it is a negative ion; if it has fewer than Z electrons, it is a positive ion.

The number of neutrons in a nucleus is represented by N, so that the *nucleon number* of a nucleus is $A = Z + N$. The nucleon number is related but not exactly equal to the atomic weight that is usually given in tables of elements. The atomic weight is a measure of the total mass of an atom (atomic mass is a better term), including electrons, which is needed for quantitative chemical calculations.

Usually (but not always) you can get the nucleon number by rounding off the atomic weight.

A given chemical element generally has nuclei with several different neutron numbers. These are called *isotopes*. For example, hydrogen has three isotopes: normal hydrogen, $_1H^1$, with no neutrons; deuterium, $_1H^2$, with one neutron; and tritium, $_1H^3$, with two neutrons. In the notation I am using here (slightly unconventional), the subscript is Z. This is technically not needed since it is already specified by the chemical symbol, but I have included it for pedagogical purposes. The superscript is A. Since all isotopes of an element have basically the same electronic structure, they will have the same chemistry except for small differences that result from their masses being slightly different.

Starting with a single proton, we add an electron and get a hydrogen atom. That electron can be found in any of the allowed energy levels described above, depending on the principle quantum number n. The lowest level, with $n = 1$, is called the *ground state*. The higher energy levels are called *excited states*. When any atom is in an excited state, it will spontaneously drop to a lower level, emitting a photon with energy equal to the energy difference between levels. It will eventually end up in the ground state. For the rest of this discussion, let us just think about ground states.

If we add another proton to the nucleus, we get helium. Since two protons have the same charge, they will repel; so we will need some neutrons—which attract protons and other neutrons by way of the nuclear force—to help keep the nucleus together. (The nuclear force will be discussed in more detail in a later chapter.) For helium, 99.999863 percent of its isotopes are comprised of $_2He^4$. The remainder occurring naturally is $_2He^3$. Isotopes of helium from $A = 5$ to $A = 10$ have been produced artificially.

Let us focus on the electrons and not worry about the number of neutrons in the nucleus, which we saw has a generally small effect on chemistry.

Recall the Pauli exclusion principle, which says that two or more fermions cannot exist in the same quantum state. Since the electron has spin ½, it has two possible spin states. So both electrons can sit in the lowest energy level.

When we add a third proton to the nucleus, we get lithium. A third electron is needed for the neutral atom. However, that electron must go into the next higher energy level, or *shell*. That is, a shell is defined as all the electrons having the same principle quantum number, n.

Recall the quantum numbers, n, ℓ, m, and m_s, that specify an electron's quantum state. As we saw above, ℓ has a maximum value of $n - 1$, m ranges from $-\ell$ to $+\ell$, and m_s is $\pm$½. So when $n = 1$, only $\ell = 0$ and $m = 0$ are possible, so only two electrons can fit in the lowest shell. When we move to $n = 2$, we can have $\ell = 0$ and $m = 0$ as well as $\ell = 1$ and $m = -1, 0, +1$ possible. There is room for eight electrons in the second shell. I leave it for the reader to work out the number of allowed electron states in the third, $n = 3$, shell. The answer is eighteen. A little mathematics will show that the maximum number of electrons in the nth shell is $2n^2$.

We also define a *subshell* as the set of states with the same orbital quantum number ℓ. The maximum number of states in the ℓth shell is $2(2\ell + 1)$.

So the periodic table is built up as electrons fill in the allowed states. Look again at figure 4.1. Note that the first row has two elements, H and He, filling up the first shell. The second row has eight, from Li to Ne, filling up the second shell. However, beyond the second shell, things get more complicated. The third row also has eight elements, and the fourth and fifth rows have eighteen, so the rows of the periodic table do not correspond simply to closed n-shells. Other effects beside the Pauli principle come into play.

For our purposes, we need not go into further technical detail. The basic point is that without the Pauli exclusion principle, the electrons in every atom would settle down to the ground state and we would have no complex chemical structures and a very different universe. It is hard to imagine that any form of life, which depends on complexity, could evolve in such a universe.

Another important point is that we can understand why some chemical elements are highly reactive, some mildly so, and some not reactive at all. It all depends on the most loosely bound electrons, the so-called *valence electrons*.

Let us examine the columns of the periodic table. In the first column, we have H, Li, Na, K, and other atoms with a single valence electron. This electron is very loosely bound and so can be easily shared with another atom. In the farthest-right column we have He, Ne, Ar, Kr, and other "noble elements" that have complete valence shells and their electrons are too tightly bound to be shared with other atoms.

Just to the left of these we have two columns of elements in which either one or two electrons are missing from a complete or "closed" valence shell. These atoms are happy to pick up an electron or two from other atoms and react with them. And so, for example, two hydrogen atoms readily combine with an oxygen atom to form H_2O. The two hydrogen electrons are shared with the oxygen atom, thus closing its valence shell.

The Schrödinger equation can be written down for any multi-electron atom, although it cannot be solved exactly for any element above helium. Nevertheless, atomic physicists and chemists can use numerical techniques to calculate many important properties of the chemical elements. All the physics needed to understand the chemical elements is contained in that one equation, and there is very little mystery left in the chemical atom.

8

INSIDE THE NUCLEUS

We knew the world would not be the same. Few
people laughed, few people cried, most people
were silent. I remembered the line from the Hindu
scripture, the Bhagavad-Gita. Vishnu is trying to
persuade the Prince that he should do his duty and
to impress him takes on his multi-armed form and
says, "Now I am become Death, the destroyer of
worlds." I suppose we all thought that, one way or
another.

—**Robert Oppenheimer**[1]

NUCLEI

Since 1932, it has been known that the nuclei of atoms are composed of protons and neutrons, generically referred to as *nucleons*. The proton carries an electric charge of $+e$, where e is the unit electric charge and the electron's charge is $-e$. The neutron is electrically neutral and is just slightly heavier than the proton. Each nucleon is over 1,830 times more massive than an electron, and so the masses of atoms are essentially contained in their nuclei.

The diameters of nuclei range from 0.8 femtometers for hydrogen to 15 femtometers for uranium, where 1 femtometer equals 10^{-15} meter. The diameter of the hydrogen atom is about 0.11 nanometer, while that of the uranium atom is 0.35 nanometer, where

1 nanometer equals 10^{-9} meter. So, roughly speaking, the nucleus of the atom is a million times smaller than the atom itself. Matter is, indeed, mostly empty space.

In chemical and nuclear reactions, the sum of the atomic numbers of the reactants does not change even though the reactants themselves can change. This follows from the principle of *charge conservation*: the total charge in a reaction is unchanged in the reaction.

Likewise, we observe a principle of *nucleon number conservation*: the total number of nucleons in a reaction is unchanged during the reaction. (Later this will be modified when we get to particle physics.)

In this chapter, all nuclear and subnuclear particles X will be designated with the notation $_Z X^A$, where Z is the charge in units of the unit electric charge (the atomic number in the case of nuclei) and A is the nucleon number, usually (but not always) the atomic weight to the nearest integer. X is the chemical symbol for the atom found in the periodic table, or the particle symbol in the case of subatomic particles.

In chapter 6, it was mentioned that Ernest Rutherford had transmuted nitrogen into oxygen by bombarding it with alpha rays. Although he did not recognize it at the time, the specific reaction was:

$$_2He^4 + {_7}N^{14} \rightarrow {_8}O^{17} + {_1}H^1.$$

Adding up the subscripts and superscripts, we see that the sums are conserved. Note that the oxygen isotope $_8O^{17}$ is produced, which has one more neutron than the most common form of oxygen, $_8O^{16}$.

THE NUCLEAR FORCES

Because the protons in a nucleus are all positively charged, they will repel one another electrostatically. Something needs to hold them together to make a stable nucleus. Gravity is far too weak for such small masses, so another attractive force is needed. This force must also apply to neutrons in order to keep them in the nucleus. In fact, the neutrons aid in keeping the nucleus together; as we add more and more protons, the mutual repulsion makes it harder to keep the nucleus together and more neutrons are needed.

The force that attracts protons and neutrons is called the *strong nuclear force* or the *strong interaction*. Experiments determine that this force is extremely short range. Unlike the gravitational and electromagnetic forces that reach across the universe, the strong nuclear force goes rapidly to zero beyond about 2.0 femtometers.

A second, even shorter-distance nuclear force exists called the *weak nuclear force* or the *weak interaction*. Recall that three types of nuclear radiation are observed: alpha, which are helium nuclei ($_2He^4$); beta, which are electrons ($_{-1}e^0$) or positrons ($_{+1}e^0$); and gamma ($_0\gamma^0$), which are high-energy photons. Alpha radiation is a manifestation of the strong nuclear force. Beta radiation results from the weak nuclear force. Gamma radiation occurs by means of the electromagnetic force. Here are examples of each:

$$\alpha \qquad _{92}U^{238} \rightarrow {}_{90}Th^{234} + {}_2He^4, \text{ and}$$
$$\beta \qquad _{55}Cs^{137} \rightarrow {}_{56}Ba^{137} + {}_{-1}e^0 + {}_0\bar{V}_e{}^0,$$

where $_0\bar{V}_e{}^0$ is an anti-electron neutrino, which will be discussed later, and

$$\gamma \qquad X^* \rightarrow X + {}_0\gamma^0,$$

where X is any generic particle and X^* is the same particle in an electromagnetically excited state. This process is analogous to the photon emission from an excited atom. Here we have photon emission from an excited nucleus, only the energy of the photon is much greater than that of an atom, MeVs rather than eVs or keVs.

"ATOMIC" ENERGY

Let us consider a generic reaction, be it chemical, nuclear, or subnuclear:

$$A + B \rightarrow C + D.$$

The number of reactants need not be two; they can be any number on either side of the arrow. And you can reverse the arrow at any time because all such reactions are reversible, although not necessarily with equal reaction rates. The principle of conservation of energy then says that the total energy before the reaction equals the total energy after, where the energy of each reactant is the sum of its rest and kinetic energies.[2]

The rest energy of a body of mass m is, of course, mc^2. If the total rest energy before the reaction exceeds the total rest energy after, an increase in kinetic energy results and the reaction produces energy, usually in the form of heat as the rise in kinetic energy corresponds to a rise in temperature. This type of reaction is called *exothermic*.

If the total rest energy before the reaction is less than the total rest energy after, a decrease in kinetic energy and cooling results. This type of reaction is called *endothermic*. If the energy difference is high enough, new particles can be produced. This does not happen in chemistry, but it is common in nuclear and particle physics. For example, since the neutron is slightly heavier than the proton, it decays producing a proton, electron, and electron antineutrino:

$$_0 n^1 \rightarrow {}_1 H^1 + {}_{-1} e^0 + {}_0 \bar{V}{}_e^0.$$

In the case of chemical reactions, the increase or decrease in kinetic energy is a tiny fraction of the rest energies of the reactants involved, and the mass differences are unmeasurable. Nevertheless, the energy produced in an exothermic chemical reaction results from the loss of rest energy in the reaction, a fact that was not appreciated until Einstein showed the equivalence of mass and rest energy, and it is still rarely explained that way in physics or chemistry classes. For example, in the chemical reaction

$$C + O_2 \rightarrow CO_2,$$

4.1 eV of energy is released, compared to the rest energy of CO_2, which is 41 GeV (1 GeV = 10^9 eV).

On the other hand, if we consider the reverse endothermic reaction

$$CO_2 \rightarrow C + O_2,$$

4.1 eV must be provided to split CO_2 into carbon and oxygen, since the rest energy of carbon dioxide is less than $C + O_2$ by that amount.

Let us define the efficiency of an exothermic reaction as the fraction of the rest energy that is converted into kinetic energy. While this definition is not what chemists or engineers generally use, it provides the precise measure of how well a reaction is able to convert its available rest energy to kinetic. For example, the efficiency of burning pure carbon to produce energy is $4.1/41,000,000,000 = 10^{-10}$. That is, for every unit of energy produced, ten billion units of the energy that is fundamentally available in principle are wasted, usually exhausted into the environment. This should drive home that the world's current use of carbon burning as its primary source of energy is incredibly inefficient.

NUCLEAR FUSION

Nuclear and particle reactions are far more efficient than chemical reactions in converting rest energy to kinetic energy. In these cases, the energies produced are a much larger fraction of the rest energies of the reactants. For example, the hydrogen isotopes *deuterium* and *tritium* will interact to produce a helium nucleus and a neutron

$$_1H^2 + {}_1H^3 \rightarrow {}_2He^4 + {}_0n^1.$$

The energy released is 17.6 MeV compared to the rest energy of $_2He^4$, which is 3,726 MeV, for an efficiency of $17.6/3,726 = 0.005$. Not great, but 50 million times better than chemical reactions. This reaction is an example of nuclear *fusion*. Basically, you can think of it as bringing two protons and two neutrons together to form a helium nucleus. Fusion is the source of energy in stars such as our sun and in the hydrogen bomb.

The primary processes that provide the energy in the sun are:

$$_1H^1 + {}_1H^1 \rightarrow {}_1H^2 + {}_{+1}e^0 + {}_0v_e^{\ 0},$$
$$_{+1}e^0 + {}_{-1}e^0 \rightarrow {}_0\gamma^0 + {}_0\gamma^0,$$
$$_1H^2 + {}_1H^1 \rightarrow {}_2He^3 + {}_0\gamma^0, \text{ and}$$
$$_2He^3 + {}_2He^3 \rightarrow {}_2He^4 + {}_1H^1 + {}_1H^1,$$

where $_{+1}e^0$ is an antielectron or positron. The second reaction is matter-antimatter annihilation. Note that the first reaction is actually a weak interaction (signaled by the neutrino), the second and third are electromagnetic interactions, and only the fourth is a strong interaction. Note also that each of the first three reactions must occur twice before the final reaction. If you put them all together, you have

$$6_1H^1 + 2(_{-1}e^0) \rightarrow {}_2He^4 + 2(_1H^1) + 2(_0v_e^{\ 0}) + 6(_0\gamma^0).$$

The net release of energy is 26 MeV.

The proton-proton interactions are very difficult to produce in most natural environments, since protons have the same charge and so repel each other. Only in the very high pressure that exists at the center of stars are the speeds of the protons high enough for them to penetrate the barrier between them. Even then, a process called *quantum tunneling* in which particles are able to tunnel through barriers, like you and I walking through a wall, is required for the reaction to take place.

It isn't magic. You and I can, *in principle*, walk through a wall by quantum tunneling. It's just that the probability for us to bounce back after hitting the wall is much higher than for us to tunnel through. At the subatomic level, on the other hand, the tunneling probability is much higher. It all depends on the thickness of the barrier.

A number of other reaction cycles are also known to take place at the center of the sun. How do we know all this about the center of the sun? We can't see it with photons. But we can with neutrinos. In 1998, I participated in an underground experiment in Japan that took a picture of the center of the sun at night using neutrinos that had not only passed through the immense solar matter lying over the sun's center, but had also passed straight through Earth as well.[3] A number of other experiments have detected neutrinos from the sun and provided detailed information about the reactions taking place.

Matter-antimatter annihilation provides 100 percent conversion efficiency from rest to kinetic energy. The problem, however, is that we have no sources of antimatter to mine. Even in the sun, the antimatter must first be generated by the initial proton-proton collision. The antimatter in the universe is a billion times less abundant than matter. We will later discuss why this is so. You would think that there should be equal amounts.

For over fifty years, attempts have been made to develop controlled nuclear-fusion energy sources, which would, in principle, provide an endless supply of energy out of water with negligible environmental impact. Unfortunately, the temperatures needed to overcome the repulsive barrier between the positively charged hydrogen nuclei are so high that they cannot be contained by any known materials. The required temperatures, on the order of 100 million degrees, are provided in a star by the tremendous gravitational pressure at its center, or, in the case of the hydrogen bomb, by a nuclear fission bomb used as a trigger. Nuclear fusion research continues with no practical application yet in sight.

NUCLEAR FISSION

When the United States annihilated the Japanese cities of Hiroshima and Nagasaki in August 1945, the world was introduced to a terrible new weapon called the "atomic bomb." This was a misnomer, since the energy of conventional explosives such as TNT results from the rapid oxidation of chemical atoms and molecules and thus is "atomic" in nature. The new source of energy was nuclear fission, and what were exploded were nuclear bombs (not "nucular").

A typical fission reaction is

$$_0n^1 + {}_{92}U^{235} \rightarrow {}_{92}U^{236} \rightarrow 4\,(_0n^1) + {}_{37}Rb^{95} + {}_{55}Cs^{137}.$$

The uranium nucleus is basically split into two nuclei, each having roughly half the mass of the uranium. The net release of energy in this reaction is 191 MeV, for an efficiency of $191/(236 \times 931) = 0.0009$. While again this is not a large number, it is 9 million times more efficient than carbon burning.

Note that three more neutrons are produced than went in, making possible a sustained *chain reaction* without which nuclear energy production on a large scale would not be possible.

The bomb makers in the World War II Manhattan Project faced a formidable problem. Fissionable ${}_{92}U^{235}$ constitutes only 0.72 percent of naturally occurring uranium, most of which is radioactive but nonfissionable ${}_{92}U^{238}$. Separating out sufficient fissionable material for a bomb required the building of a huge facility at Oak Ridge, Tennessee. After an excruciatingly long time of centrifuging enough pure $_{92}U^{235}$, a uranium bomb could be built. Nicknamed "Little Boy," it was dropped over Hiroshima on August 6, 1945.

Also part of the Manhattan Project, another huge facility was built in Hanford, Washington. There, so-called breeder reactors transformed $_{92}U^{238}$ into fissionable isotopes of plutonium, of which only trace amounts are found in nature. While several plutonium isotopes are fissionable, the primary isotope used in plutonium bombs is $_{94}Pu^{239}$ that is produced by the reactions

$$_{92}U^{238} + {_0}n^1 \rightarrow {_{92}}U^{239} \rightarrow {_{93}}Np^{239} + {_{-1}}e^0 + {_0}\bar{\nu}_e{}^0, \text{ and}$$
$$_{93}Np^{239} \rightarrow {_{94}}Pu^{239} + {_{-1}}e^0 + {_0}\bar{\nu}_e{}^0.$$

When enough plutonium was finally accumulated, it was used to build a plutonium bomb named "Fat Man." It was dropped over Nagasaki on August 9, 1945.

Following World War II, there existed much enthusiasm that nuclear energy would solve the world's energy problems. There also existed much dread that humans would destroy the world with uncontrolled nuclear war. Neither have yet to come to pass.

Mutually assured destruction (MAD) seems to have kept another nuclear bomb from being used against an enemy since the tragedies of Hiroshima and Nagasaki. The end of the Cold War makes large-scale nuclear war increasingly unlikely, while at the same time, nuclear proliferation makes small-scale nuclear conflicts and terrorist attacks a serious threat.

POISONING THE ATMOSPHERE

Currently more than 80 percent of the world's energy is obtained from fossil fuels: petroleum, coal, and natural gas. Another 10 percent comes from biomass and waste, leaving only 10 percent from sources other than carbon burning. The CO_2 and other compounds produced by these fuels are filling the atmosphere with pollutants. The World Health Organization estimates that urban air pollution results in 1.3 million premature deaths worldwide each year. Indoor air pollution is estimated to cause approximately 2 million premature deaths, mostly in developing countries. Almost half of these deaths are due to pneumonia in children under five years of age.[4]

Observations unequivocally show that carbon dioxide levels in the atmosphere have risen to levels higher and faster than any on record and that Earth is definitely warming by way of the greenhouse effect. The overwhelming majority of climate scientists agree that this warming will result in severe climate change in a few years, if it hasn't already begun, and that the burning of fossil fuels by human industry is the primary cause.

In my previous book, *God and the Folly of Faith*, I described how powerful corporations with economic interests in fossil fuels have used their vast funds to foment groundless doubt about these solid scientific results. As part of their strategy to preserve their freedom to exploit resources for profit, they have allied themselves with the antiscience Religious Right, which believes God created Earth's resources for us, the pinnacle of Creation, to consume. And so, the majority of evangelical Christians, including prominent Republican politicians, believe that God will assure that Earth remains in perfect environmental balance—at least until Jesus returns.[5]

The sad part of the story is that with nuclear technology that was developed over sixty years ago, we could have preserved a clean atmosphere with safe levels of pollutants. The happy part of the story is, we can still do it.

NUCLEAR POWER

As of December 2011, thirty countries worldwide were operating 435 nuclear reactors for generating electricity, and 63 new nuclear plants were under construction in fourteen countries. Nuclear power now provides 13.5 percent of the world's electricity. France is the largest percentage user, with 74.1 percent of its electricity coming from nuclear reactors. In 2010, the United States obtained 19.6 percent of its electricity from nuclear reactors.[6]

However, nuclear energy has not proved to be "too cheap to meter," as the chair of the Atomic Energy Commission, Lewis Strauss, had enthused in 1954.[7] The technologies involved turned out to be much tougher, and nuclear power plants are far more costly than originally imagined. Radioactive waste disposal is a major issue. Still, the environmental hazards of this waste dwarf in comparison to those from carbon burning, which as we have seen results in over a million deaths each year. Despite this fact, nuclear power today is feared by much of the populace.

On March 28, 1979, a series of equipment malfunctions and operator errors led to a partial meltdown of the nuclear reactor at Three Mile Island in Pennsylvania. While negligible radiation was released, and no one was injured, the event caused widespread panic and triggered a moratorium on nuclear power plant construction in the United States that has continued to this day. Not a single new plant was approved until 2012, with two recently being approved. The last plant to go into operation in the United States was one ordered in 1976.

A far more serious nuclear disaster occurred at Chernobyl in Ukraine on April 26, 1986. Once again, operator errors were a major contributing cause and explosions blew the top off a reactor and set off a fire. Radioactive clouds spread over a large portion of the continent. A 2005 report by the International Atomic Energy Agency, the World Health Organization, and the United Nations Development Agency estimated that up to four thousand people could ultimately die of radiation exposure from the accident—bad, but still orders of magnitude less than the deaths from carbon burning.[8]

Chernobyl resulted in an even more intense opposition to nuclear energy, especially in Europe. Only France continued its program at full speed, while Germany began phasing its program out. Beyond Europe, Japan lacked any significant energy sources and, despite being the only nation ever to have suffered a nuclear attack, continued building nuclear power plants. Dozens more came online after Chernobyl. By the beginning of 2011, Japan had fifty-four operating plants supplying 30 percent of its energy.[9] But then, on March 11 of that year, nature triggered a disaster of immense proportions that once again affected the whole nuclear picture.

On that fateful day, an earthquake just off the northeastern coast of Honshu generated an immense tsunami that swept ashore, destroying everything in its path. Unfortunately, that path included the Fukushima Daiichi nuclear station. While the radiation released was not as bad as Chernobyl, it was bad enough, and another blow was struck against the use of nuclear energy.

Many lessons were learned from the mistakes that resulted in the three nuclear plant accidents. Designs were poor and certainly could have been improved, especially with regard to safety. It surely was a major error in judgment to locate the Fukushima station so close to a fault line. While both Three Mile Island and Chernobyl exhibited design flaws, the actual equipment failures were minor. However, plant operators made wrong decisions that reflected inadequate training that ended up doing more harm than good. Put simply, the standards were simply too low and the designs were inadequate.[10]

But, as Pulitzer Prize–winning author Daniel Yergin says in his in-depth study of the economics and politics of world energy, nuclear power remains "the only large-scale, well-established, broadly deployable source of electric generation currently available that is carbon-free."[11] Despite the fact that the United States has not put a new nuclear power plant into operation for decades, during which time power consumption in the country has more than doubled, nuclear energy still provides 20 percent of America's electricity needs, showing that nuclear technology continues to improve.

LIQUID FLUORIDE THORIUM REACTORS

Unfortunately, all existing nuclear power plants utilize the fission of $_{92}U^{235}$ and $_{94}Pu^{239}$. This was a terrible choice that was based on making bombs and powering submarines and other ships rather than achieving optimal energy production. An alternative that has been available from the beginning is the fission of $_{92}U^{233}$. While $_{92}U^{233}$ does not occur naturally, it can be bred from $_{90}Th^{232}$ (thorium), which is four times as abundant in nature as $_{92}U^{238}$, by the reactions

$$_{0}n^1 + {}_{90}Th^{232} \rightarrow {}_{90}Th^{233} \rightarrow {}_{91}Pa^{233} + {}_{-1}e^0 + {}_{0}\overline{V}{}_{e}^{\ 0}, \text{ and}$$
$$_{91}Pa^{233} \rightarrow {}_{92}U^{233} + {}_{-1}e^0 + {}_{0}\overline{V}{}_{e}^{\ 0},$$

where Pa is protactinium. The same initial reaction produces other reaction products

$$_{0}n^1 + {}_{90}Th^{232} \rightarrow {}_{90}Th^{231} + 2{}_{0}n^1,$$
$$_{90}Th^{231} \rightarrow {}_{91}Pa^{231} + {}_{-1}e^0 + {}_{0}\overline{V}{}_{e}^{\ 0}, \text{ and}$$
$$_{0}n^1 + {}_{91}Pa^{231} \rightarrow {}_{91}Pa^{232} \rightarrow {}_{92}U^{232} + {}_{-1}e^0 + {}_{0}\overline{V}{}_{e}^{\ 0}.$$

In particular, $_{92}U^{232}$ is nasty because gamma rays with energies of 2.6 MeV are radiated in its decay chain. This makes $_{92}U^{233}$ undesirable for weapons use because a weapon should not radiate much until it is detonated so that it needs only minimal shielding. Gamma rays are dangerous to the personnel who have to handle the weapon and to the weapon's instrumentation. However, these gamma rays are not a factor for use in reactors, which are already sufficiently shielded.

Fission bombs were based on $_{92}U^{235}$ or $_{94}Pu^{239}$, where the gamma radiation was minimal and of lower energy. So when the time came to build nuclear reactors for power production, including propulsion for navy ships and submarines, this was where the knowledge and expertise was developed. While seven different designs of nuclear power plants are currently in commercial operation, all are based on uranium or internally bred plutonium, with Japan and Russia each having one plant that utilizes plutonium alone. The most common design is the pressurized-water reactor, developed by Alvin Weinberg at Oak Ridge in 1946, based on the earlier designs of Enrico Fermi in Chicago. These comprise 265 of the 439 reactors in use.

Unfortunately, reactors based on $_{92}U^{235}$ and $_{94}Pu^{239}$ have designs and other inherent flaws that contributed to the nuclear power plant disasters that turned the public, and many scientists, against nuclear power:

- They require costly separation of uranium isotopes or breeding of plutonium.
- Most use solid fuels that must be removed before they have given off 1 percent of their potential energy availability, because of depletion of the fissionable material and radiation damage.
- Water or other coolants under high pressure must be pumped through from the outside to extract heat. Meltdown can occur when coolant flow is disrupted, as with power failures.
- High pressure requires high-strength piping and pressure vessels that are degraded by radiation and corrosion. If weakened, the reactor vessel or parts of the pressure boundary can rupture, causing a loss of cooling and the potential release of radioactivity to the environment.
- Even after shutdown, fission products continue to generate heat and must be cooled with water for one to three years. If this flow is disrupted, meltdown can occur.

- The radioactive waste takes 10 million years before it decays to naturally occurring levels.
- Proliferation. Material that can be used in bombs can be secretly extracted from these reactors.
- Large size and high cost of facilities.
- Like oil, coal, and natural gas, uranium probably will run out in less than a hundred years unless large numbers of fast-neutron-breeder reactors are built. These types of nuclear reactors have significant safety issues, and prototypes generally have not been successful. It seems doubtful that uranium can provide a long-term solution to the world's energy needs.

During the same postwar years, Weinberg was also responsible for another reactor design based on ideas from Fermi and Eugene Wigner. His design was successfully developed at Oak Ridge but never put into commercial use. This was the liquid-fuel reactor in which the fuel is composed of molten salt.

I will just discuss one version of the molten-salt reactor in which thorium is used as the input fuel, the thorium fluoride reactor (LFTR, referred to as "Lifter").[12] This utilizes the reaction described above in which $_{90}Th^{232}$ is transmuted to fissionable $_{92}U^{233}$. After successful tests of the molten-salt concept at Oak Ridge, the LFTR project was canceled in 1969 by the Nixon administration. The reason: having two fewer neutrons than $_{92}U^{235}$, $_{92}U^{233}$ is not as efficient in breeding plutonium for bombs. Instead, a liquid-metal fast-breeder $_{92}U^{235}$ reactor that could efficiently breed plutonium was funded.

Today there is renewed worldwide interest in LFTR, although the United States continues to drag its feet, spending very little on research. While many scientific and engineering problems still have to be solved, LFTR seems sufficiently promising that it might, in fact, be able to provide for the increasing needs for energy, especially in developing countries, while finally reducing our reliance on fossil fuels. I will just list the advantages. Some—but not all—apply to other types of liquid-fuel reactors.[13]

- Thorium is plentiful and inexpensive. One ton, costing $300,000, can power a 1,000 megawatt plant for a year.
- LFTR operates at atmospheric pressure, obviating the need for a large, expensive containment dome and having no danger of explosion.
- It cannot melt down because the normal operating state is already molten.
- Stable to rising temperatures because molten salt expands, slowing the reaction. A salt plug kept solid by cooling coils will automatically melt if external power is lost; the fluid drains out to a safe dump tank.
- Salts used are solid below 300 F, so any spilled fuel solidifies instead of escaping into the environment.
- Liquid fuels use almost all the available energy, unlike solid fuels that must be removed before they have generated 1–3 percent of the available energy.
- The radiative waste is much less than from conventional plants and far more manageable.
- Air-cooling is possible where water is scarce.
- Thorium can be cheaper than coal.
- Proliferation resistant. It can't be used to build bombs.
- Are far easier to scale to meet a wide range of energy needs.
- Could provide for the world's energy needs, carbon-free, for a thousand years.

If the Three Mile Island, Chernobyl, and Fukushima reactors had been LFTRs, no radiation would have been released to the environment. Indeed, no meltdown or explosion would have occurred.

While three hundred years for the decay of radiative waste to natural levels still seems undesirable, it must be put in perspective. Even the waste from conventional nuclear plants is tiny compared to the waste from burning fossil fuels, which, although not (as) radioactive, still kills millions of people each year. A great effort is going on now to develop carbon sequestration, in which the CO_2 from coal power plants is pumped back into the ground. This will be expensive, raising the price of electricity from coal an estimated 80–100 percent.[14] Furthermore, the amount of underground space needed to store a year's worth of CO_2 output from a single coal power plant is equivalent to six hundred football fields filled to a height of ten yards. By comparison, one football field filled to the same height is required for all the waste from the entire civilian nuclear program.[15]

In short, while unforeseen problems could still arise, it would seem that a major effort in the United States and worldwide should be undertaken to develop nuclear power systems based on liquid fluoride thorium. For a good summary of the world energy problem and the role LFTR can play, see the online lecture by physicist Robert Hargraves.[16] After this section was written, an excellent new book appeared titled *Superfuel: Thorium, the Green Energy Source for the Future,* by science writer Richard Martin, that gives the full story of thorium.[17]

9

QUANTUM FIELDS

*Every instrument that has been designed to be
sensitive enough to detect weak light has always
ended up discovering the same thing: light is made
of particles.*

—**Richard Feynman**[1]

PHYSICS IN 1945

The great physicists of the early twentieth century gave us relativity, gave us quantum mechanics, and revealed the structure
of atoms and nuclei. Their successors were poised to carry on with
their work when World War II intervened and the younger generation was instead put to work designing bombsights, sonar and radar
detection, instrument flight, and, of course, the nuclear bomb. At the
same time, mathematicians were breaking German and Japanese
codes while chemists, geologists, and other physical scientists were
doing their part in putting science to use in the war effort.

They all succeeded spectacularly so that by war's end the physical sciences and mathematics were viewed in a new light by the
public and politicians alike. Soon biology joined in the spotlight as
its contributions to medical science, spurred on by the new technologies developed during and after the war, began to have a growing
impact on human health. Science was respected as never before—
and as never since.

Our concern in this book is atomism, so let us review the status of the physics of matter as of 1945, when the Manhattan Project scientists could get back to doing what they really wanted to do with their time—basic research.

By that period, all of known matter could be built from protons, neutrons, and electrons. Light was composed of photons, so it seemed that all we needed in order to understand everything was four "elementary" particles. Unfortunately (or fortunately for physicist employment), these four particles weren't the only ones out there. In 1932, the same year that Chadwick discovered the neutron, Carl D. Anderson found the antielectron, or positron, whose existence had been predicted by Dirac in 1929. In 1936, Anderson and Seth Neddermeyer stumbled upon another particle in cosmic rays that best can be described as a heavy electron. We now call it the *muon*. It is 207 times more massive than the electron, but it is similar in every other respect, including having an antimatter partner comparable to the positron.

And this wasn't all. Beta-decay reactions led Pauli to propose in 1930 that a neutral particle with small or possibly zero mass was carrying off invisible energy. Enrico Fermi dubbed this hypothetical particle the *neutrino*, the "little neutral one." Experimental confirmation of the neutrino did not occur, however, until 1956, and it was eventually discovered that three different types of neutrinos and their antiparticles existed.

Despite these empirical successes, and many more that were to occur rapidly after the war, no credible models describe how particles interacted with one another. Four fundamental forces could be identified:

- The *gravitational force*, which acts on all particles including, as Einstein showed, the photon.
- The *electromagnetic force*, which acts on all electrically charged particles plus the photon.
- The *strong nuclear force*, which acts on only protons and neutrons.

- The *weak nuclear force*, which acts on all particles except photons.

Although Einstein's general theory of relativity is not a quantum theory, it accounts for all observations involving gravity. What's more, gravitational effects are observable only when one of the bodies involved is very massive, such as Earth or the sun, and are negligible otherwise. And so, in 1945, quantum physicists decided to focus on electromagnetism—in particular, the interaction of photons and electrons.

MORE HYDROGEN SURPRISES

Recall from chapter 6 that the relativistic quantum theory developed by Paul Dirac in 1928 was able to successfully calculate the fine structure of the spectral lines of hydrogen. These were attributed to the interaction between the magnetic field produced by the electron in its orbit around the nucleus and the intrinsic magnetic field of the electron itself. Most importantly, Dirac's equation showed that the magnetic moment of the electron, a measure of the strength of the field, is twice what it should be if the electron were simply a classical spinning charged sphere, just as had been observed. Furthermore, the spin quantum number of the electron was ½, which made no sense if it were just the sum of the orbital angular momenta of particles inside the electron, which have integral quantum numbers. From all that was known at the time, and all that is known to date, the electron is a point particle with no substructure. Points don't spin. Quantum spin has no classical analogue.

According to the Dirac theory, the atomic states with $n = 2$ and $\ell = 0$ or 1 (see chapter 6) have the same energy, 3.4 eV. That is, they are "degenerate," and they sure look that way in the optical region of the spectrum. In 1947, Willis Lamb and Robert Retherford at Columbia University measured the microwave radio emissions from hydrogen and discovered that these two hydrogen states were separated by 1,058 million cycles per second (mc, now called mega-hertz), a frequency corresponding to an energy of 4.3 millionths of an electron volt. This was named the *Lamb shift*.

At the same time at Columbia, Isidor Rabi (1898–1988) and his students John Nafe and Edward Nelson observed a second split in a spectral line. In this case, the 13.6 eV ground state of hydrogen ($n = 1$, $\ell = 0$) was observed to be split by 1,421.3 mc, or 5.8 millionths of an eV.

This so-called hyperfine splitting was perhaps less of a surprise than the Lamb shift and attributed to the interaction between the magnetic field of the electron and that of the proton nucleus. In 1934, Rabi and collaborators at Columbia had measured the magnetic moment of the proton and found a value larger than expected for a pure "Dirac particle" such as the electron. Using the theoretical Dirac value of the electron's magnetic moment and the measured Rabi value for the proton, a splitting of $1,416.90 \pm 0.54$ mc was calculated. While this was close, the discrepancy was nevertheless significant given the exquisite precision of the measurements involved.

The fact that the measured magnetic moment of the proton was not what it should have been for an elementary particle was the first evidence that the proton is not a point particle but that it has a substructure. Furthermore, the neutron, which has zero charge, has a nonzero magnetic moment showing that it, also, is not elementary. In fact, the neutron's magnetic moment is negative, as if it contained a negative charge circulating about a positive charge (like an atom!). Later, as we will see in the next chapter, the proton and neutron are composed of fractionally charged quarks.

QED

Just prior to World War II, Dirac and several of the other quantum pioneers had developed a relativistic quantum field theory based on his relativistic quantum mechanics. This turned out to be calculable only to a first-order approximation. When the physicists tried to move to the next order, they encountered infinities in their calculations.

In 1947, Hans Bethe performed a calculation of the Lamb shift in which he treated the atom nonrelativistically and absorbed the infinities into the mass and charge of the electron, a process called *renormalization*. He obtained 1,040 mc, in good (if not perfect) agreement with the measured 1,062 mc.[2]

Following up on Bethe's intuition, a fully relativistic theory of how photons and electrons interact, renormalizable to all orders of approximation, was developed independently by Julian Schwinger[3] and Richard Feynman[4] in the United States and by Sin-Itiro Tomonaga[5] working in war-ravaged Japan. The theory was dubbed *quantum electrodynamics* or QED.[6] In 1949, Freeman Dyson, a Brit working at Princeton, showed that the three formulations were equivalent.[7]

In the spring of 1948, Feynman attended an invitation-only meeting at the Pocono Manor Inn in rural Pennsylvania. There he introduced a simple line drawing, now called a *Feynman diagram*, which would prove to be an invaluable tool in particle physics.[8]

The Feynman diagram was originally motivated by the need to keep track of various terms in QED calculations. Although Schwinger and Tomonaga did not use them, Feynman diagrams are very intuitive visual models of fundamental interactions that supplement a cookbook set of rules for calculating interaction probabilities. While many physicists view them as just calculational tools, others, like myself, look at them as bearing some relation to what is actually going on in nature.[9]

The first-order interaction of two electrons is illustrated in figure 9.1. One of the electrons (solid lines) emits a photon (dashed line), which carries energy and momentum to the other electron.

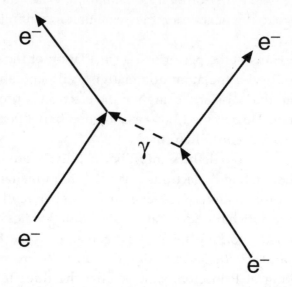

Figure 9.1. First-order Feynman diagram for the interaction between two electrons (e⁻) by way of the exchange of a photon (γ). The electron coming up from the right emits a photon, which travels through space and is then absorbed by the electron coming up from the left.

The lines in the diagram can be thought of as roughly representing the paths of particles in a two-dimensional space-time picture, with a space-axis to the right and a time axis up. Or, you can think of the arrows as representing momentum vectors. In neither case is this visual aid drawn to scale.

The notion that particles interact by means of exchanging other particles was not original with Feynman. Hans Bethe and Enrico Fermi had already proposed the idea of photon exchange in 1932. And in 1942, Ernst Stückelberg had used space-time diagrams similar to Feynman's to describe electron-positron pair production and annihilation.[10] In any case, with Feynman diagrams and the rules for their use, we had a fully relativistic prescription for calculating the probability to higher orders of approximation.

If an electron is in an atom or in the presence of some external field, we can view the first-order interaction as in figure 9.2 (a), where the source of the incoming photon, not shown, is either the electric field or the magnetic field of the nucleus. To first order, the field emits a single photon. Feynman diagrams for second-order interactions are shown in figure 9.2 (b), (c), and (d). These provide the main contribution to the anomalous magnetic moment of the electron that deviates from the Dirac value. They're not as complicated as they look at a first glance. Just follow the particle lines from bottom to top and you can see the sequence of interactions.

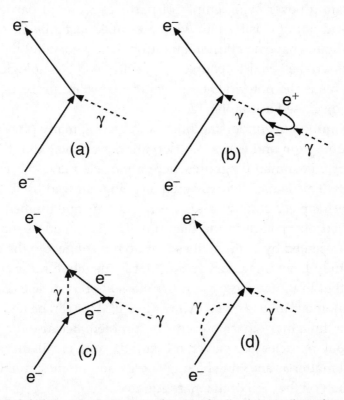

Figure 9.2. (a) First-order Feynman diagram for the interaction of an electron in a hydrogen atom with an external electromagnetic field that can be either the electric field (Lamb shift) or magnetic field of the nucleus; (b) second-order diagram with electron-positron loop; (c) second-order diagram with "triangle" loop; (d) second-order diagram in which an electron emits and reabsorbs a photon.

Figure 9.2 (b), in which a photon converts to an electron-positron pair, which then converts back to a photon, is an example of *vacuum polarization*. During the time that pair is present, an electric dipole (pair of positive and negative charges) that can affect the path of the electron exists in otherwise empty space. This is the primary source of the Lamb shift, where, we can think crudely, the electron has different orbits for $\ell = 0$ and $\ell = 1$, so the two states are affected differently by the polarized vacuum.

We can also see from this discussion why physicists talk about the vacuum never being empty of particles. Pairs of particles can flit in and out of existence in what are called *quantum fluctuations*. Isolated pairs have no effect in quantum field theory, but these fluctuations are believed to couple to gravity as a cosmological constant. This is the notorious cosmological-constant problem, which will be discussed in chapter 12.

A common misunderstanding, even among many physicists, is that momentum and energy conservation are violated at the "vertices" of a Feynman diagram where a particle may split into two particles. This is usually explained as being allowed by the uncertainty principle. Actually, this is wrong. Energy and momentum are conserved everyplace in the diagram. The so-called virtual particles represented by internal lines that do not connect to the outside world may have imaginary mass, that is, negative mass squared. Recall that in units where $c = 1$, the mass m of a particle of energy E and momentum p is given by $m^2 = E^2 - p^2$. This can be negative or positive. Imaginary mass presents no problem because the masses of virtual particles are never measured. And recall that I have defined matter as anything having energy and momentum, with no restriction on the sign of the mass squared.

In 1948, Feynman showed that an antiparticle going one direction in time is empirically indistinguishable from the corresponding particle going backward in time.[11] The idea also had been proposed by Stückelberg in 1942.[12] So, in all the diagrams we can draw, we can always reverse the arrow and change an electron to a positron.

In the case of the photon, you are also free to reverse its arrow, but it is its own antiparticle, so there is no difference.

We see that many different diagrams can be drawn. However, note that they all have the same basic reaction in which two electrons interact with a photon at a point called a *vertex*. As shown in figure 9.3 (a), we can view it as an electron being scattered by a photon, or in figure 9.3 (b), as an incoming photon becoming an electron-positron pair. Reverse other lines and you will have other interpretations that are all experimentally indistinguishable.

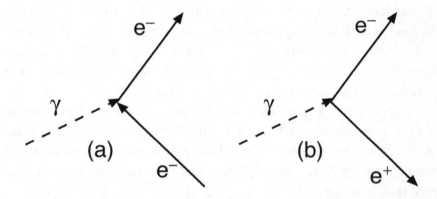

Figure 9.3. The basic QED interaction vertex. In (a), an incoming electron is deflected by an incoming photon. In (b), the incoming electron's arrow is reversed so it becomes an outgoing positron, and the incoming photon creates and electron-positron pair.

In 1948, using his own highly sophisticated mathematical techniques, not based on Feynman diagrams (and so incomprehensible except to a select few), Schwinger calculated the electron magnetic moment by taking into account the second-order corrections. He obtained a value that was greater than the Dirac value by a fractional amount of 0.00118, in good agreement with experiment. Since then, QED calculations for the electron magnetic moment up to fourth order agree with the most precise measurements to ten significant figures.

Schwinger also obtained 1,040 mc for the Lamb shift, the same value calculated by Bethe. The current QED calculation, including higher orders, is 1,058 mc, which is in exact agreement with the current value of 1,057.864 mc. Since this original work, quantum electrodynamics has been tested to great precision without failure. QED may be the most precise theory in all of science.

FIELDS AND PARTICLES

The extraordinary success of quantum electrodynamics suggests that relativistic quantum field theory is telling us something about the nature of reality. But what exactly is it telling us? To explore that question, let us avoid the complications of interacting particles and consider an electromagnetic field all by itself.

Suppose we have an array of harmonic oscillators, for example, simple pendulums, each with a different natural frequency f. An electromagnetic field is mathematically equivalent to such an array, where each oscillator has unit mass and corresponds to one "radiation mode" of the field.[13]

In quantum mechanics, a harmonic oscillator of frequency f has a series of ladderlike, equally spaced energy levels separated by energy hf, where h is Planck's constant. Because of the uncertainty principle, an oscillator can never be at rest and the lowest energy level is $\frac{1}{2}hf$. This is called the *zero-point energy*.

It follows, then, that the energy of an electromagnetic field is quantized in steps of hf, as proposed by Planck. When a transition occurs from one energy level to the next one below, a quantum of energy hf is emitted that we interpret as a photon of that energy. Thus, the field can be thought of as containing a certain number of photons, each of energy hf. Starting at the bottom of the ladder, the zero-point energy state has zero photons. The next rung of the ladder has one photon; the next, two; and so on. The total energy of each radiation mode of the field is then the zero-point energy

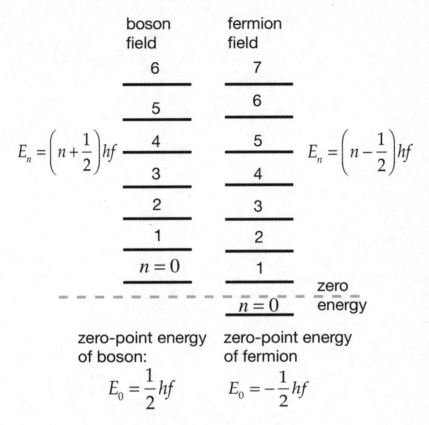

Figure 9.4. The energy levels of a free boson and fermion field.

plus the sum of the energies of the photons in the field: $\frac{1}{2}hf + hf + 2hf + 3\,hf + \ldots\,$.

Now, this picture can be shown to apply to all bosons (integer spin particles). In the case of fermions (half-integer spin particles), we have the same ladder except the zero-point energy is $-\frac{1}{2}hf$. (Negative energies are allowed in relativity.)

In short, a *field-particle unity* exists such that for every quantum field, there is an associated particle that is the *quantum* of the field. In the 1970s, physicists developed the standard model of particles and forces based on relativistic quantum field theory in which each elementary particle is the quantum of a quantum field.

10

THE RISE OF
PARTICLE PHYSICS

*You don't need something to get something more.
That's what emergence means. Life can emerge from
physics and chemistry plus a lot of accidents. The
human mind can arise from neurobiology and a lot
of accidents, the way the chemical bond arises from
physics and certain accidents. It doesn't diminish
the importance of these subjects to know they follow
from more fundamental things plus accidents.*
                              —Murray Gell-Mann[1]

PION EXCHANGE AND THE STRONG FORCE

With the incredible success of quantum electrodynamics, it was natural to apply the same methods of relativistic quantum field theory to the nuclear forces. Recall from chapter 8 that in 1932 Hans Bethe and Enrico Fermi had proposed that particles interact by exchanging other particles. In 1934, physicist Hideki Yukawa suggested that two nucleons (protons or neutrons) interact by the exchange of a particle called the *meson*. Note that this was well before the photon-exchange mechanism was applied with such great success in quantum electrodynamics.

Yukawa's model specified that the mass of an exchanged particle is inversely proportional to the range of the force that is mediated by that particle.[2] Because the photon is massless, the electromagnetic force has unlimited range, and so light can reach us from great distances. Because the range of the strong nuclear force is only about a femtometer (10^{-15} meter), Yukawa estimated the mass of the meson to be about two hundred times heavier than the electron but still much less massive than the proton.[3]

In 1936, Carl D. Anderson discovered the particle we now call the *muon* in cosmic rays. This particle was produced by the primary cosmic rays hitting the top of the atmosphere. At first, it was thought that the observed particle was Yukawa's meson, since its mass was 207 times the mass of the electron, almost exactly as predicted. However, it was soon realized that the muon was not strongly interacting and so could not be responsible for the nuclear force. Any strongly interacting particles produced in the upper atmosphere would quickly lose most of their energy colliding with the nuclei of atoms in the atmosphere. By contrast, muons easily reach sea level and penetrate deep into the ground. One muon passes through every square centimeter of our bodies every minute. As previously mentioned, the muon is essentially a heavy electron.

In 1947, a better candidate for Yukawa's meson was found in cosmic rays at high altitude. This particle is now called the pi meson or *pion*. The pion is strongly interacting, and few reach sea level. The pion occurs in three varieties: π^+ with charge +1, π^- with charge –1, and π^0 with charge 0. The mass of the π^0 is 135 MeV, while the mass of each charged pion is 140 MeV. These masses actually fit the Yukawa model better than the muon mass. Pion exchange makes possible the variety of nucleon-nucleon interactions shown in figure 10.1.

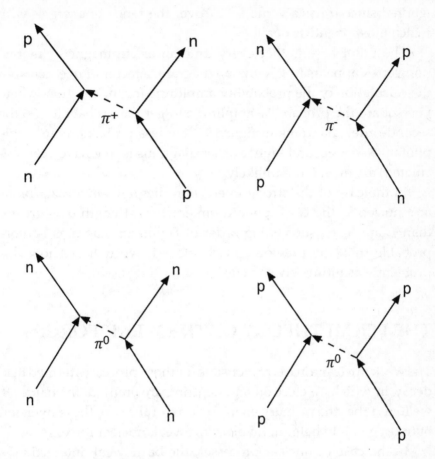

Figure 10.1. Yukawa's proposal that nucleons interact by meson exchange.

However, all attempts to construct a QED-like theory for the strong interaction, with the pion taking the place of the photon, were a failure. I should know. I spent several years trying to make the model work. The reason photon exchange is such a good first approximation to the electromagnetic interaction between electrons is that the force strength is weak. The electromagnetic force strength is given by a dimensionless quantity $\alpha = e^2/\hbar c$ called, for historical reasons, the *fine-structure constant*, where e is the unit electric charge. Although it has a slow dependence on energy, this is generally ignored since α has a value $1/137$ over the range of energies with which most scientists deal.

The point here is that every time an electromagnetic interaction vertex appears in a Feynman diagram, a factor of α^2 enters into the calculation of the probability amplitude for the reaction, where you square the probability amplitude to get the probability. So the second-order diagrams in figure 9.2 are less probable than single photon exchange, and higher-order diagrams (some have been calculated) are even more unlikely.

In the case of the strong-interaction diagrams involving pions and nucleons, the corresponding interaction strength α_s is greater than 1, and each succeeding order of Feynman diagrams is more probable than the previous. In chapter 11, we will see how this problem was ultimately solved.

THE FERMI THEORY OF THE WEAK FORCE

The weak nuclear force is responsible for those processes termed beta decay, in which an electron and neutrino are emitted. In chapter 8, we found that the primary energy source of the sun, the conversion of hydrogen into helium, involves the weak nuclear force.

In the case of nuclear processes, the basic weak interaction is neutron decay,

$$_0n^1 \rightarrow {}_1p^1 + {}_{-1}e^0 + {}_0\overline{\nu}{}_e^0$$

($_1p^1$ is the same as $_1H^1$). Nuclear beta decay occurs when a neutron inside a nucleus decays. Most nuclei are stable, however, because their binding energies prevent their decay products from escaping the nucleus. Radioactive nuclei are the exceptions.

In 1934, Fermi developed a model of beta decay that was based on the diagram shown in figure 10.2.

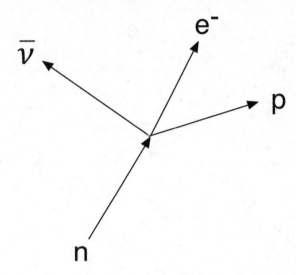

Figure 10.2. Fermi model of beta decay.

Note that the interaction occurs at a point, implying that the range of the force is zero.

The Fermi model worked fairly well and was improved on over the years. However, the weak force was unlikely to have exactly zero range, and it seemed reasonable to assume an exchange process, such as illustrated in figure 10.3, where the exchanged particle W is a postulated *weak boson* that mediates the weak interaction. The weak interaction strength α_W is very small, so single W exchange suffices.

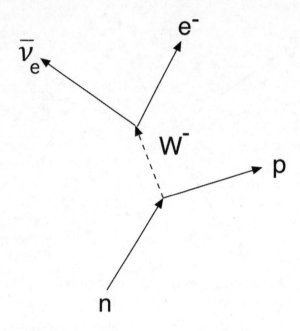

Figure 10. 3. Neutron beta decay by way of the exchange of a weak boson W.

However, searches for a weak boson in the 1960s came up with nothing. This was not unexpected because the range of the weak force is even shorter than the strong force and so the W was probably too massive to be produced with the existing accelerators of the day. But its day would come.

THE PARTICLE EXPLOSION

Besides the muon and pion, cosmic-ray experiments in the 1950s revealed several additional particles whose behavior was strange, and so they were called *strange particles*. These included four

K-mesons or *kaons*: the charged K$^+$ and K$^-$, each with a mass of 494 MeV and two neutral kaons, K^0, $\bar{\text{K}}^0$, each with a mass of 498 MeV. In addition, three types of particles heavier than nucleons called *hyperons* were found. These strange particles were assigned a new quantum number called *strangeness*, which appeared to be conserved in strong interactions but not in weak interactions. For example, charged kaons were produced in oppositely charged pairs in cosmic rays, never singly, which was explained by assigning the two particles of the pair opposite strangeness. Strangeness was assumed to be conserved in the interaction that produced the pairs. (See further discussion of strangeness conservation below.)

In the 1960s, new accelerators at the Lawrence Radiation Laboratory in Berkeley, California (now the Lawrence Berkeley Laboratory), the Brookhaven National Laboratory on Long Island, and the CERN laboratory in Geneva (now the European Center for Particle Physics) went into operation. They produced hundreds of new particles that were clearly not composed of other known particles. I participated in these events as a graduate student at the University of California at Los Angeles (UCLA) and then, after graduating in 1963, as an assistant professor of physics at the University of Hawaii where two colleagues and I set up a particle-physics research group, which still thrives today.

Almost all the new particles were strongly interacting and called *hadrons*. The exceptions were a second heavy electron, the *tauon*, which joined the electron and muon in a non-strongly interacting class called *leptons*. All three had charge –1. Included in the lepton class were three types of neutrinos, one associated with each charged lepton: the *electron neutrino, muon neutrino*, and *tauon neutrino*. Each lepton had an antiparticle partner.

The heavy electrons and all the new hadrons were very short-lived and did not occur in nature except momentarily when produced in the upper atmosphere by high-energy cosmic-ray collisions. However, the cosmic-ray "beam" was not controllable, and accelerators provided a much more efficient source and a better ability to examine the particle properties. The primary device used to detect these particles was the now-obsolete *bubble chamber*, although other instruments contributed as well.

Hadrons were divided into two categories: *baryons*, such as the proton and neutron, that had half-integral spin; and *mesons*, such as the pions and K-mesons, that had zero or integral spin. I did my doctoral thesis at UCLA on the interaction of K^+ mesons, interacting with deuterium (heavy hydrogen) in a bubble chamber located at the Bevatron accelerator in Berkeley.

With a bubble chamber, we could photograph the paths of charged particles by the trails of bubbles they left in a superheated fluid, which produced beautiful images that enabled us to observe and measure the momenta and energies of the particles produced in a reaction. In the next chapter, we will review the attempts to bring order to this chaos of new particles, and how by the 1970s particle physicists had succeeded spectacularly with what was modestly called the *standard model of particles and forces*.

First, however, I need to cover another two important developments that occurred in the late 1950s: (1) the observation of several new conservation principles and (2) the discovery of broken symmetries in the weak interaction.

NEW CONSERVATION PRINCIPLES

Reactions involving the new particles were found empirically to obey a set of new conservation rules in addition to the familiar ones—energy, momentum, angular momentum, electric change, and nucleon number.

Recall from chapter 8 that the number of nucleons, that is, protons and neutrons, in a nuclear reaction is the same before and after the reaction takes place. This was no longer the case for the new particles. Instead, it was found that the number of baryons was conserved. To show this, let me generalize the notation introduced in chapter 8 so that a particle X is denoted by $_Z X^B$, where Z is the electric charge in units of the proton charge e and B is the *baryon number*. Both Z and B are conserved in all reactions observed so far.

For example,

$$_0\pi^0 + {}_1p^1 \rightarrow {}_1K^0 + {}_0\Lambda^1,$$

where $_0\pi^0$ is the neutral pion, $_1p^1$ is the proton, $_1K^0$ is the K-meson with charge $+e$, and $_0\Lambda^1$ is the electrically neutral lambda hyperon.

The proton, neutron, and hyperons have B = +1, while their corresponding antiparticles have B = −1. The mesons and leptons have B = 0.

This example also serves to illustrate the conservation of strangeness, S, mentioned previously. The pion and proton have S = 0, the K-meson has S = +1, and the lambda has S = −1, so strangeness is conserved in the above reaction.

On the other hand, reactions such as

$$_0\Lambda^1 \rightarrow {}_1p^1 + {}_{-1}\pi^0$$

are observed, where $_{-1}\pi^0$ is the negatively charged pion. This violates strangeness conservation. However, this is a weak interaction, whereas the previous is a strong interaction. We conclude that strangeness is conserved in strong interactions but not in weak interactions. Note that charge and baryon number are still conserved.

While other conservation principles were discovered as more particles were produced at higher energies, I will mention just one more here—lepton number conservation.

Recall that three negatively charged leptons—the electron, the muon, and the tauon—were found, and each was associated with a neutrino. These have been assigned a lepton number L = −1, while their corresponding antiparticles have L = +1. The hadrons (baryons and mesons) have L = 0.

Consider neutron beta decay,

$$_{0}n^{1} \rightarrow \ _{1}p^{1} + \ _{-1}e^{0} + \ _{0}\bar{\nu}_{e}^{0}.$$

This exhibits charge, baryon number, and lepton number conservation.

While no violation of baryon number or lepton number conservation has been observed, as we will see in chapter 12 both must have been violated in the early universe to account for the large excess (by a factor of a billion) that exists for matter over antimatter.

BROKEN SYMMETRIES

Not everything we witness in the world is symmetrical. While we often think of Earth as a sphere, it does not possess perfect spherical symmetry. That is, Earth is not the same viewed from all angles. Earth's rotation causes it to bulge at the equator and flatten at the poles, which breaks spherical symmetry and makes Earth an oblate spheroid. On the other hand, it is roughly symmetric about its axis of rotation. We say that Earth has axial symmetry but broken spherical symmetry.

Similarly, when we look in a mirror, the face we see is not the same one others see when they look straight at us. Our faces break left-right or mirror symmetry. We might call that "broken mirror symmetry. "

Broken symmetry is a feature of the familiar phase transitions we experience in everyday life. When water vapor condenses into liquid water and liquid water freezes into ice, each transition is a broken symmetry. Similarly, when a magnetite, a form of iron oxide that is naturally magnetic, is very hot, it will be nonmagnetic. Then, when it is cooled below a critical temperature, it becomes magnetic. Say the magnetite is shaped in a sphere. Above the critical temperature, it has spherical symmetry, which is then broken at lower temperature as the direction of the magnetic field singles out a special direction in space.

Until the 1950s, it was thought that fundamental chemical, nuclear, and elementary particle processes possess mirror symmetry, technically known as *parity* symmetry. That is, a reaction always seemed to occur at the same rate as the equivalent reaction viewed in a mirror. As you know, right and left are interchanged in a mirror. The parity operation exchanges left and right and is designated by the letter P.

In 1956, puzzling observations involving kaons in cosmic rays led to the suggestion by two young Chinese physicists working in the United States, Tsung Dao Lee and Chen Ning Yang, that parity symmetry was broken in weak interactions. The violation of parity symmetry was observed in the beta decay of cobalt-60 nuclei in 1956 by a team led by another Chinese physicist in the United States, Madame Chen-Shiung Wu.

The operation of changing a particle to an antiparticle, originally called *charge-conjugation*, is designated by the letter C. Until 1964 the combined operation CP, where you exchange particles with their antiparticles and left with right, seemed to be invariant for all chemical, nuclear, and particle reactions. However, that year CP symmetry was found to be violated in neutral kaon decays by Princeton physicists James Cronin and Val Fitch.

A symmetry that can be shown to follow from the axioms of quantum field theory is CPT, which combines C, P, and the time-reversal operation, T. Thus, the violation of CP leads to the conclusion that T is violated as well. In 1998, evidence for direct T violation independent of CP was found in neutral kaon decays.

However, as mentioned in chapter 5, it is incorrect to conclude that the source of the arrow of time, and thus the second law of thermodynamics, lies in these discoveries. As we have seen, the arrow of time is a statistical definition that applies for systems of many particles. While, as I have emphasized, so-called irreversible events on the macroscopic scale are still in principle reversible, the probability for the reverse process is so small as to be unlikely to occur in the age of the universe.

By contrast, the kaon reactions on the submicroscopic scale we are talking about here can occur in either direction. The violation of time-reversal symmetry does not mean time reversal is impossible. It just means that the rate in one direction is different than the rate in the other. In the case of kaon decays, the difference is typically only one part in a thousand.

In any case, these observations of a small amount of symmetry breaking are limited to weak interactions. You can take any reaction and, according to our best knowledge, predict that the CPT inverse will occur at the same rate.

"NUCLEAR DEMOCRACY" AND *THE TAO OF PHYSICS*

There is one more story I need to cover before we get to the standard model. As we have seen, in the 1960s accelerators of ever-increasing energies were producing hoards of new, short-lived particles that were clearly not composed of electrons and nucleons and could not all be "elementary." Quantum field theory was having a lot of trouble explaining them and began to be questioned. An attempt was made to do away with the idea of elementary particles altogether and replace it with a new "bootstrap" theory in which all the particles are somehow made up of each other. The leading proponent was a brilliant, Hollywood-handsome professor from the University of California at Berkeley named Geoffrey Chew. He called the notion "nuclear democracy." In this picture, there are no elementary particles. Or, if you wish, they are all elementary.

This was not as crazy as it sounds. Consider figure 10.4, where the arrows on incoming and outgoing nucleons in the π^+-exchange diagram in figure 10.1 have been reversed. Recall that an antiparticle going one direction in time is empirically indistinguishable from the corresponding particle going backward in time. In figure 10.4 (a), we have an antineutron colliding with a proton, producing a pion, which then decays back into an antineutron and a proton. So, in a sense, the pion is sometimes a nucleon-antinucleon pair.

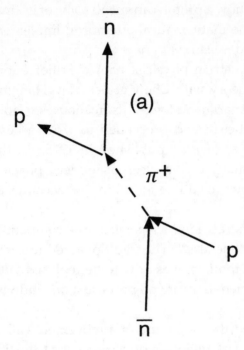

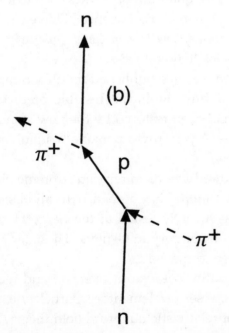

Figure 10.4. In (a), an antineutron and a proton collide to make a pion, which then decays back to an antineutron and a proton. In (b), a pion and a neutron collide to produce a proton, which then decays back to a pion and a neutron. These are examples of how particles can be viewed as sometimes being composed of other particles.

In figure 10.4 (b), we can see how a proton can spend some of its time as a pion and a neutron. The mathematical attempt to implement the bootstrap idea was called *S-matrix theory*.[4]

In the early 1970s, an Austrian physicist named Fritjof Capra was working on S-matrix theory with Chew at Berkeley. Hanging around Berkeley's famous marijuana-fogged coffeehouses, Capra learned about Eastern mysticism and proceeded to write a best-selling book called *The Tao of Physics*, published in 1975.[5] In the book, Capra claimed that many of the ideas of modern physics, especially quantum mechanics, could be found in the teachings of Eastern mysticism.

Capra viewed the bootstrap idea as particularly compatible with the mystical notion of "oneness."[6] The traditional reductionist physics, as exemplified by atomism, was now to be replaced with a new holistic physics in which there are no parts, just one indivisible whole.

The Tao of Physics marked the beginning of a movement called the "New Age" in which, in part, quantum physics is used to justify a new kind spirituality ostensibly based on science.[7] I have covered this story in detail in two books, *The Unconscious Quantum*[8] and *Quantum Gods*.[9] Here I will stick to the physics.

By the time *The Tao of Physics* was published in 1975, S-matrix theory had failed to produce any significant testable predictions while reductionist physics had been restored by the new standard model of particles and forces. Again, I was personally involved in that effort.

Once again, atomism and reductionism reigned supreme. Since its introduction, the standard model has agreed with all observations and is only now being severely tested at the Large Hadron Collider (LHC) at the CERN laboratory in Geneva. These tests are unlikely to overthrow atoms and the void.

Not only did the standard model restore atomism and reductionism by introducing a new set of elementary particles, it also reestablished the integrity of relativistic quantum field theory. The standard model provided a completely relativistic, renormalizable scheme in which equations can be written down that fully describe all that is known about the fundamental particles and forces of nature. As we will see, the key ingredient was the application of a symmetry principle called *gauge invariance*.

11

THE DREAMS THAT STUFF IS MADE OF

The effort to understand the universe is one of the very few things which lifts human life a little above the level of farce and gives it some of the grace of tragedy.

—Steven Weinberg[1]

THE QUARKS

Considerable theoretical effort went into attempting to classify the new particles that were turning up at the accelerator labs. In 1961, something akin to the periodic table of the chemical elements was independently developed by a Caltech physics professor Murray Gell-Mann and Israeli physicist (and high-ranking military officer and government official) Yuval Ne'eman. It was called the *Eightfold Way*, after the Buddha's path to enlightenment. More technically, the theory was based on a mathematical symmetry group called SU(3). Symmetry groups are mathematical classifications of various types of symmetry transformations. For example, the group of rotations in three dimensions is O(3).

In 1964, Gell-Mann and Caltech graduate student George Zweig independently proposed a natural explanation for the Eightfold Way in which hadrons, including the proton and neutron, were composed of more elementary constituents that Gell-Mann dubbed *quarks*. Until this time, all known particles had either zero, positive, or negative electric charge that was an integer multiple of the unit electric charge, *e*. The electron has charge –*e*; the proton has charge +*e*. Gell-Mann and Zweig made the crucial assumption that quarks had fractional charge. Using current terminology, the proton is composed of two u ("up") quarks, each with charge +2/3*e*, and a d ("down") quark with charge –1/3*e*, for a net charge of *e*: uud. The neutron is udd for a net charge of zero. Note there is no negative nucleon that can be formed with a triad of u and d quarks.

Antiprotons and antineutrons, which had been confirmed at the Berkeley Bevatron in 1955 and 1956, are made of antiquarks. The antiproton, $\bar{u}\bar{u}\bar{d}$, has charge –*e*. The antineutron $\bar{u}\bar{d}\bar{d}$ is electrically neutral.

Baryons in general are composed of three quarks. Their antiparticles are composed of three antiquarks. Mesons, on the other hand, are composed of quark-antiquark pairs.

Gell-Mann and Zweig also proposed a third quark, now called s ("strange"), which has charge –1/3*e* and a strangeness $S = -1$. Its antiparticle $\bar{s}$ has charge +1/3*e* and a strangeness $S = +1$.

The same year, a particle called the omega-minus baryon (Ω^-), which had previously been predicted by Gell-Mann and others, was found after a surprisingly short search at Brookhaven National Laboratory.[2] It had exactly the properties expected if it was composed of three s quarks (sss). That is, it is a singlet of charge –*e* and strangeness –3. The discovery bubble-chamber picture and its interpretation is given in figure 11.1. No Ω^0 or Ω^+ was ever seen, consistent with the fact that sss is the only combination of three s quarks. An antimatter version $\bar{\Omega}^+$ with charge +*e* and strangeness +3 was detected shortly thereafter. With the discovery of the omega baryons, the quark model of matter was convincingly established. It was further reinforced by the failure to discover any hadrons that could not be composed of quarks.

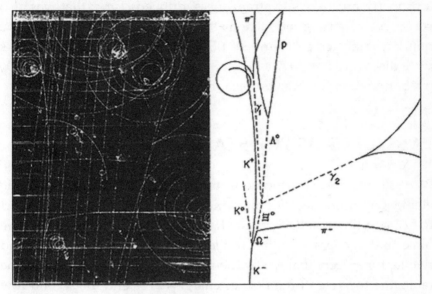

Figure 11.1. The bubble-chamber picture of the first omega-minus. An incoming K-meson interacts with a proton in the liquid hydrogen of the bubble chamber and produces an omega-minus, a K^0 meson, and a K^+ meson, which all decay into other particles. Neutral particles, which produce no tracks in the chamber, are shown by dashed lines. The presence and properties of the neutral particles are established by analysis of the tracks of their charged decay products and application of the laws of conservation of momentum, energy, and charge.

Experiments in the 1970s in which electrons and neutrinos were scattered from protons and neutrons (I worked on several of the neutrino experiments at Fermilab and CERN) revealed point-like structures inside the nucleons with fractional charge, fitting all the hypothesized properties of quarks. These experiments provided direct evidence for point-like constituents inside the proton and neutron, just as the Rutherford-Geiger-Marsden experiment provided direct evidence for the point-like nucleus inside the chemical atom.

The three-quark model successfully described all the hadrons up until 1974, when evidence was found for a fourth quark with a property analogous to strangeness, dubbed *charm*, that was also conserved in strong interactions but not in weak interactions. The quark was labeled c ("charmed"). Soon hadrons containing c quarks were discovered. In 1977 evidence was found for yet another quark, b ("bottom" or, sometimes, "beauty").

PARTICLES OF THE STANDARD MODEL

By 1980, the particles shown in table 11.1 were established as the elementary ingredients of the standard model that remains in effect today, pending results from the Large Hadron Collider (LHC).[3] We have three generations, or families, each having two quarks and two leptons, accompanied by their antiparticles (not shown). In the table, the top row of quarks has charge $+2/3e$ and the second row has charge $-1/3e$. All the hadrons that had been discovered by the end of the 1980s were built from combinations of these quarks. The t ("top" or sometimes, "truth") quark that joins with the b quark in the third generation is so much heavier than the other quarks (174 GeV, where 1 GeV = 1 billion electron volts, and a proton mass is 0.938 GeV) that it was not confirmed until 1995, when Fermilab reached sufficiently high energy to produce it.

The top row of leptons in table 11.1 contains the three different types of neutrinos, each having zero charge. The second row has the electron, muon, and tauon, with charge $-e$. The antiparticles have opposite charge. The masses of the particles are also shown in the table.

The story of neutrino mass is a profound one. For years, it appeared that the masses of the neutrinos were all zero. However, there was no theoretical reason for this to be the case. If they had masses, however, experiments already established that these masses were far smaller than the lightest known particle with mass, the electron.

The first evidence for a nonzero mass for neutrinos came in 1998 in an experiment called Super-Kamiokande, in a mine in Japan.[4] I played a minor role in this experiment; however, in 1980 I proposed the technique that was used at a neutrino conference in Wisconsin in the context of another planned experiment that was never constructed.[5]

The neutrinos listed in the table, v_e, v_μ, v_τ, do not have definite mass states, which means they are not stable. Rather they are mixtures of three stable neutrino mass states v_1, v_2, v_3 that have the following mass (square) limits,

$$\left| m_2^2 - m_1^2 \right| \approx \left(9 \times 10^{-3} \right)^2 \text{ eV}^2 \text{ and } \left| m_3^2 - m_2^2 \right| \approx \left(4 \times 10^{-2} \right)^2 \text{ eV}^2.$$

Neutrinos have a remarkable, purely quantum mechanical property called *neutrino oscillation*. If you start out with a pure beam of muon neutrinos, for example, it will eventually transmogrify into a beam containing all three types. This was the property my collaborators and I at Super-Kamiokande observed in 1998, which determined that neutrinos have mass. If they were massless, they wouldn't oscillate.

The evidence that neutrinos have mass came from the observations of the flux of muon neutrinos produced in the upper atmosphere as they passed through Earth to the detector. Those traveling a greater distance had the greatest chance of changing type and, sure enough, we saw fewer muon neutrinos coming straight up through the center of Earth than just below the horizon.

The quarks and leptons are all fermions with spin ½. In addition, the standard model contains twelve "gauge bosons," spin 1 particles that mediate the various interactions between the quarks and leptons. These include the photon, which mediates the electromagnetic interaction as described previously. Three bosons, W^+, W^-, and Z, mediate the weak interaction. Eight *gluons*, g, mediate the strong interaction. Some examples of elementary particle reactions are given in figure 11.2.

Fermions (antiparticles not shown)			Bosons
Quarks	u 2.3 MeV	c 1.27 GeV	t 173 GeV	γ 0
	d 4.8 MeV	s 95 MeV	b 4.18 GeV	g 0
Leptons	ν_e *see text*	ν_μ *see text*	ν_τ *see text*	Z 90.8 GeV
	e 0.511 MeV	μ 106 MeV	τ 1.78 GeV	W 80.4 GeV

Table 11.1. The fundamental particles of the standard model and their masses. The Higgs boson is not shown.

While this may seem complicated, remember that these elementary particles are all we have needed for the last three decades to account for every observation without a single anomaly. This contrasts with the anomalies with the phenomenon of light that we have seen existed at the close of the nineteenth century. Only now, with the LHC, are new elementary particles to be anticipated.

Indeed, most of these elementary particles are of interest only to particle physicists and cosmologists. When considering the ingredients of matter that most people deal with—not only in everyday life, but also in almost all scientific laboratories everywhere—the number of elementary particles needed now is the same as was needed in 1932. Recall that in 1932 all of matter was seen to be composed of protons, neutrons, and electrons. Along with the photon to provide light, all we needed was four particles. Today, while the LHC data are being accumulated, most of us can still get by with four particles: the u and d quarks, the electron, and the photon.

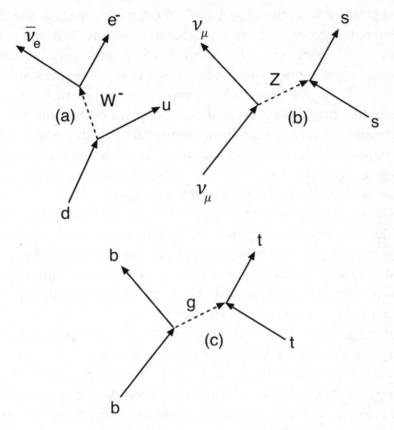

Figure 11.2. Examples of interactions in the standard model. Beta decay is shown as the decay of a d quark in (a). A muon neutrino interaction with an s quark by way of Z exchange is shown in (b). The strong interaction between a b quark and a t quark by gluon exchange is shown in (c).

GAUGE SYMMETRY

Early in the twentieth century, another symmetry was discovered that played a role of increasing importance and eventually provided the foundation of the standard model of elementary particles and forces. This was *gauge symmetry*. The easiest way I can think to describe gauge symmetry is to view it as rotational symmetry inside the abstract, multidimensional space that physicists use to provide additional degrees of freedom beyond those provided by four-dimensional space-time. These degrees of freedom include electric charge along with the various quantities that appear in the standard model, such as baryon number, lepton number, strangeness, and charm. In string theory, the additional degrees of freedom appear as curled-up extra-spatial dimensions, but my discussion does not require this assumption. Indeed, I am deliberately avoiding covering string theory in this book because I want to stick to only well-established, empirically based knowledge.

Gauge symmetry first appeared in classical electrodynamics. Besides the conservation principles already discussed, another conservation principle has been a part of physics since the first experiments on electricity by Benjamin Franklin and others. Although not immediately recognized as such, this is the principle of charge conservation that I introduced in chapter 8.

If charge is conserved, then there must be a corresponding symmetry principle.[6] This principle is gauge symmetry.

Classical electrodynamics contains a four-dimensional vector field called the *vector potential* from which the electric and magnetic fields can be calculated. It was discovered long ago that it was possible to change this field by what was called a *gauge transformation* without changing the electric and magnetic fields.[7] It was proved that invariance under a gauge transformation leads to charge conservation.

Earlier, I likened a gauge transformation to a rotation of a coordinate system in an abstract space. In the case of quantum mechanics, the state of a system is specified by a *state vector* in such a space. In the simplest case, the space is the two-dimensional complex plane and the state vector is represented by the wave function, which is a complex number having a magnitude and a phase.[8] A rotation in the complex plane changes the phase but not the magnitude of the wave function, so the probability of the state, given by the square of the magnitude, is unchanged. That is, the probability calculated in quantum mechanics is invariant to the rotation of the complex plane, as illustrated in figure 11.3. The invariance of probability under gauge transformations is called *unitarity*.

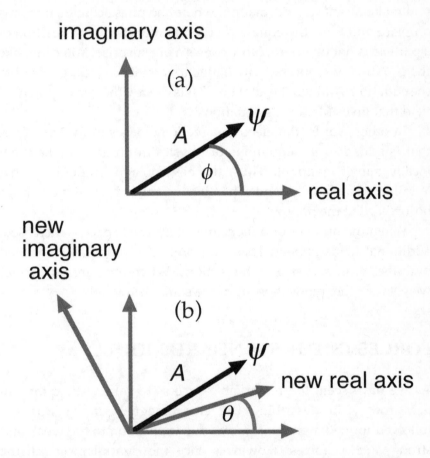

Figure 11.3. The state vector ψ in the two-dimensional complex plane. In (a), its magnitude A and phase ϕ are shown. In (b), the coordinate system has been rotated by an angle θ. The state vector is invariant to the rotation. Note, however, that while the vector does not change in magnitude or direction under the rotation, the phase does change because it depends on the choice of axes.

When you make a *global* gauge transformation, where the phase change is the same at every place in space and time, the equation of motion for the wave function of a particle is invariant to the change in phase, which leads to charge conservation. However, if you make a *local* transformation, where the phase change depends on place and time, the equation of motion for the wave function of a particle is not invariant. Now comes a big surprise. You can make the equation of motion locally gauge invariant by adding a field to the equation. And what is that field? It is exactly the vector potential field that gives Maxwell's equations!

In other words, the electromagnetic field is simply an artifact that is added to the equations of physics in order to make them locally gauge invariant. Thus, the electromagnetic field is what is called a gauge field, and when it is quantized, we get a "gauge boson" called the photon.

When the number of dimensions of abstract space is increased, additional fields appear. These are precisely the fields associated with the weak and strong interactions. When they are quantized, we get the other gauge bosons of the standard model.

FORCES IN THE STANDARD MODEL

As we have seen, the remarkable success of relativistic quantum field theory in describing the electromagnetic force was not followed immediately by a successful application to the weak and strong nuclear forces. However, once the quark-lepton scheme of elementary particles was established and the power of gauge transformations recognized, theoretical physicists were able, in a few short years in the 1970s, to produce a relativistic quantum field theory that unified the electromagnetic and weak forces into a single entity called the *electroweak force*. Abdus Salam, Sheldon Glashow, and Steven Weinberg shared the 1979 Nobel Prize in Physics for this work.

Electroweak unification was also a theory of the breaking of electroweak unification. That is, it showed how it itself is violated. The electromagnetic and weak forces are unified at very high energies, above about 1 TeV (1 TeV = 1 trillion electron volts). At these energies, the masses of the W and Z bosons are zero, like the mass of the photon. But this leads to the weak force having the same unlimited range as the electromagnetic force. This grossly contradicted the data from the sub-TeV experiments of the time, which indicated a range of the weak force a thousand times smaller than the radius of a proton.

Weinberg and Salam independently discovered a process by which electroweak symmetry is broken. This is called the *Higgs mechanism* and will be described in the next section. Associated with the Higgs mechanism is a particle called the Higgs boson. As we will see later in this chapter, a particle of mass in the range 125 to 126 GeV that looks very much like the Higgs boson was reported by two independent experiments at the LHC on July 4, 2012.

The theory of electroweak symmetry breaking predicted that the mass of the charged W bosons would be 80 GeV and the mass of the neutral Z boson would be 91 GeV, where in these units the proton mass is 0.938 GeV. The W and Z bosons at exactly these masses were detected months apart in 1983 by two independent experiments at CERN. This discovery was considered so important that the experimental leaders, Carlo Rubbia and Simon van der Meer, were awarded the Nobel Prize in Physics with unprecedented speed, just a few months after the announcement. For a very nice, up-to-date history of particle physics that covers all these events up until the end of 2011, just as the Higgs data were coming out, see *Massive: The Missing Particle That Sparked the Greatest Hunt in Science* by science journalist Ian Sample of the *Guardian*.[9]

During the same time that the electroweak theory was being developed, physicists working on the strong interaction were able to develop a relativistic quantum field theory of the strong interaction. Dubbed *quantum chromodynamics* (QCD), it became part of the standard model. QCD was based on the idea that quarks carry what is called *color charge*. Quarks are proposed to come in three "colors," red, green, and blue, where these properties are analogous to those of the familiar primary colors of visible light, although there is no direct connection between the two phenomena.[10]

All observed hadrons are color neutral, that is, "white." Baryons, such as the proton and omega-minus, are composed of one red, one green, and one blue quark. By being different colors, the Pauli exclusion principle, which prevents identical fermions from being in the same quantum state, is avoided.

Antiquarks have the corresponding complementary colors: cyan, magenta, and yellow. Mesons, such as the pion and kaon, are composed of one quark and an antiquark of complementary color, so they are also white.

In QCD, the strong force is mediated by gluons. The basic interaction vertex is seen in figure 11.4.

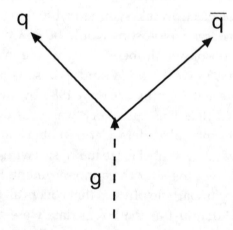

Figure 11.4. The strong interaction vertex in QCD. A gluon produces a quark-antiquark pair.

There are eight different gluons: two white (colorless) and six with all the colorful combinations that can be made from red, green, and blue quarks and their cyan, magenta, and yellow antiquarks.

For example, red plus cyan is white, while red plus magenta is something nonwhite.

Tests of QCD have been difficult because no one has ever produced free quarks in the laboratory. Indeed, if one did, that would constitute a violation of one of the prime tenets of QCD: free quarks cannot exist. The reason is that the color force increases with the separation between two quarks. When that separation exceeds a few femtometers, the color charge is discharged. This is analogous to the discharge that takes place between two electrically charged bodies in a conducting medium that we see as a spark or a bolt of lightning. While a vacuum is nonconducting for electric charge, it is conducting for color charge.

Recall that the single pion exchange diagram failed as a mechanism for the interaction of nucleons because of the high value of the strength factor α_s. So why should we expect it to work here? Good fortune. Recall that the electromagnetic strength factor α varied very slowly with energy, but too slowly to worry about for most applications. In the case of α_s, the energy dependence is rapid, and at high energies α_s becomes small enough to weaken the force. This is called *asymptotic freedom*, since quarks become increasingly free as they collide at higher and higher energies.

But you still can't kick them out of a nucleus (at least according to QCD) because of color charge breakdown. When you bang quarks together at increasingly high energies, as in a proton-proton or proton-antiproton collider, you just produce more and more quark-antiquark pairs that form mesons and quark-quark-quark trios that form baryons. And that's what you see when you look at the pictures taken at today's colliding beam detectors: $E = mc^2$ in all its glory. The more energy, the more new particles are made from that energy.

All the aspects of the standard model have been subject to numerous tests, and no experiment has, as of this writing, uncovered any violation. However, the real crucial tests are coming up as the LHC moves into full-scale operation.

THE HIGGS BOSON

The unification of the electromagnetic and weak forces just described was strongly counterintuitive. If the two forces are just two different manifestations of the same basic electroweak force, how can their ranges of influence be so vastly different? The electromagnetic force has unlimited range, reaching us from galaxies 13 billion light-years away, while the weak nuclear force can reach only 10^{-18} meter, a thousand times less than the radius of a proton. And these are the same force?

As we have seen, the bosons that act as the carriers of forces have masses inversely proportional to their ranges. The photon has zero mass, while the weak bosons, W and Z, have masses of 80 GeV and 91 GeV, respectively. The strong force is an exception. The gluon has zero mass and the small range of the strong force results from other factors. As described in the preceding section, the vacuum conducts color charge, which prevents gluons from traveling more than about a proton radius.

Clearly, electroweak unification is broken at our current level of experimentation. This is an example of spontaneous broken symmetry.

The standard model is based on gauge symmetry. That symmetry is believed to hold at very high temperatures, such as those in the early universe. At that time, the W and Z bosons and all other particles had zero mass. So a fundamental question is, how did these particles and others get their masses?

In 1964, a decade before the development of the standard model, six physicists published three independent papers in the same issue of *Physical Review Letters* suggesting a way by which particles gain mass.[11] To his expressed embarrassment, the process was named the *Higgs mechanism* after just one of the authors, a British professor named Peter Higgs. The other authors deserve mention. They are: Robert Brout (now deceased), François Englert, Gerry Guralnik, Dick Hagen, and Tom Kibble.

I referred earlier to the very informative and entertaining history of particle physics published in 1993 by Nobel laureate and then Fermilab director Leon Lederman, with help from Dick Teresi, called *The God Particle*.[12] Lederman had used that expression to describe the Higgs particle in a lecture, partly as a joke. As we have seen, he is a great comedian. However, Lederman was also partly serious in emphasizing the fundamental role of the Higgs particle, existing everywhere in space, sort of like a god, and providing for the masses of particles. Most physicists hated the name. Higgs winced at it, calling it aggrandizing and offensive. Lederman said he offended two groups of people: those who believe in God and those who don't. But his publisher liked it and the media loved it.[13]

Higgs and the other authors had shown mathematically how the weak bosons are able to gain mass by a phase transition, the spontaneous breaking of the underlying gauge symmetry. The process left behind a field of scalar (spin-zero) particles called Higgs bosons that exist throughout space.[14]

Let me try to explain the concept in simple terms, although, as always, one can never do justice to a theory when words have to be substituted for mathematics.

In chapter 12, I will discuss the cosmology of the early universe. For our purposes here, let us assume it was originally in a state of perfect symmetry. As an analogy not to be taken too literally, think of a spherical ball of magnetite, a naturally magnetic form of iron oxide, heated to a very high temperature. Because of thermal motion, the iron atoms in the ball, which are like little bar magnets, point in random directions so, as a whole, the system is spherically symmetric. As the magnetite cools, it drops below a certain critical temperature and the random motions of the atoms slow down, allowing the little magnets to line up with each other because of their mutual interaction. The result is that the ball of magnetite becomes a magnet. Spherical symmetry is broken as a particular north–south axis is singled out. Note that this direction is random. This is an example of spontaneous symmetry breaking. The breaking of the symmetry results in the appearance of a field—the magnetic field.

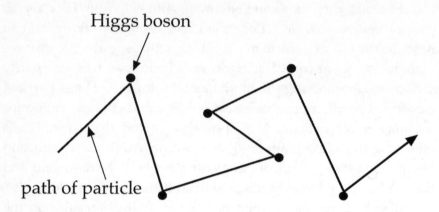

Figure 11.5. How a massless particle gains mass in the Higgs field. The particle scatters off Higgs bosons in the field, causing the particle to move more slowly through space, sluggishly, as if it had mass.

Now, in the case of the Higgs mechanism, the symmetry that is broken is not in space-time but abstract state-vector space. Fields that are generated in this way are called *gauge fields*. The Higgs field is a gauge field. And the Higgs boson is the quantum of the Higgs field.

So how does the Higgs boson generate the masses of elementary particles? A crude, intuitive explanation is given in figure 11.5.

As a particle passes through the medium of the Higgs field, it collides with the Higgs bosons that manifest the field. Although the particle is moving at the speed of light along each leg, it scatters off the Higgs bosons and effectively slows down its progress through the medium. That is, it behaves more sluggishly—as if it had more inertia or mass. The photon does not interact directly with the Higgs and so remains massless.

Although, as we have seen, the Higgs mechanism was suggested several years earlier; it was built into the standard model in the 1970s. For over a decade, the Higgs boson remained the only particle in the model whose existence was not empirically confirmed—that is, until July 2012. Note it is the only spin-zero particle in the standard model, although there may be additional Higgs bosons in extensions of that model that might come about as new data come in.

The Higgs boson turned out to be the key to electroweak unification. As mentioned, Weinberg and Salam discovered that the Higgs field was responsible for the symmetry breaking that separates electromagnetism from the weak nuclear force leaving the photons massless while the W and Z bosons gain mass. And that wasn't all the Higgs did for the theory. In 1971, Dutch physicists Gerhardus 't Hooft and Martinus Veltman proved that the Higgs mechanism provides for the renormalization of all theories of the type called *Yang-Mills* theories, of which the electroweak force and other forces of the standard model are specific examples.

MAKING AND DETECTING THE HIGGS

Let me give an idea of how the Higgs boson might be produced and detected at the Large Hadron Collider. The LHC collides protons again protons.

In figure 11.6, one way of many that this can produce Higgs bosons is illustrated. Gluons emitted by quarks in the protons interact to make a Higgs boson.

Now what about detection? Again many possibilities exist, but I will look at the two that have provided the best current evidence. In figure 11.7 (a), we have the Higgs decaying into two photons (γ-rays). In (b), the Higgs decays into two charged lepton pairs (ℓ) that can be either muons or electrons.

In either case, the energy and momentum of the outgoing particle assumed to be from the Higgs decay are measured. From these, the mass of whatever produced the observed particles can be calculated. If that object is a particle, this should show up as a peak in the statistical distribution of as many such events as can be accumulated. The greater the number of events, the greater the statistical significance of the peak. When the chances of the peak being a statistical fluctuation are less than about one in ten thousand (in particle physics convention; some other fields have embarrassingly lower standards), a tentative discovery can be claimed. Only after independent replication, however, will the scientific consensus express confidence in the claim.

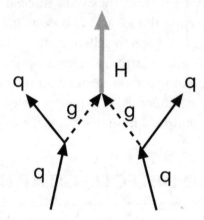

Figure 11.6. Gluons emitted by quarks in each of the proton beams interact to make a Higgs boson.

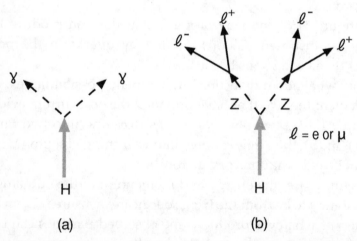

Figure 11.7. In (a), a Higgs decays into two photons. In (b), a Higgs decays into two Z bosons that then decay into charged lepton pairs (ℓ). The leptons can be either electrons or muons.

HUNTING THE HIGGS

Despite being proposed almost fifty years ago at this writing, and despite its crucial role in the highly successful standard model, the Higgs boson remained until now the only unconfirmed ingredient of that model. This has been presumably because its mass was too high to be produced with the particle accelerators available prior to the LHC. Ian Sample provides a detailed survey of the various experiments that unsuccessfully sought evidence for the Higgs until the first hints seen at the LHC in December 2011.[15] I will provide only a sketch.

In the 1980s, US physicists set out to go full speed ahead and build the largest colliding beam accelerator the world had ever seen: the *Superconducting Super Collider* (SSC). Although at the time I was working on very high-energy gamma-ray and neutrino astronomy, I attended a couple of the summer workshops that Fermilab organized in Snowmass, Colorado. I was still an accelerator physicist at heart and supported the SSC enthusiastically.

President Ronald Reagan gave his approval, and in 1988 the site was selected to be Waxahachie, Texas. It may have been no coincidence that the new president elected that year, George H. W. Bush, was (nominally) a Texan. However, geologically it was a good site and the state of Texas also put up $1 billion to help pay costs. Construction began in 1991. The collider was to be a ring 87.1 kilometers (54.1 miles) around and would collide beams of protons together, each with energy 20 TeV.

Lederman's book, *The God Particle*, along with another excellent popular book by Steven Weinberg, *Dreams of a Final Theory*,[16] both published about the same time, were intended to generate public support for the project, which would cost billions. When, in 1993, the estimated cost jumped from $4.4 billion to $12 billion, Congress took another look. By that time, $2 billion had already been spent and the tunnel for the ring was almost 30 percent complete.

However, the project had strong opposition from influential

physicists in other fields, notably Nobel laureate Phillip Anderson, who deeply resented the way particle physicists were always saying their work was more fundamental than his field of condensed matter physics.[17] Yet the Higgs mechanism was first suggested by Anderson in his work on superconductors a year before the Higgs et al. publications. Furthermore, the accelerator would involve the most massive use of superconductivity, for which Anderson won his award, the world had ever seen.

Also voicing strong opposition was the incoming president of the American Physical Society, materials scientist James Krumhansl. With this high-level opposition within the physics community, and with a Congress unconvinced of its value, the SSC didn't stand a chance.[18]

Weinberg, who teaches at the University of Texas, feels that the SSC might have had better luck if the originally favored site, Lederman's Fermilab, had been chosen.[19] Much of the infrastructure and expert staff would have already been in place rather than having to be built up from scratch. With the scientifically unjustified $25 billion International Space Station also earmarked for Texas, the two giant Texas projects had little support in the rest of the country. The SSC was canceled by a vote of 2 to 1 in Congress, while the space station passed by one vote. Although President Bill Clinton had voiced mild support for the SSC, he signed the bill canceling the project on October 31, 1993.

This event marked a sharp decline in support for elementary particle physics in the United States that has continued to this day. Physics graduate students (including my own son) looked elsewhere for opportunities. (He found a good one in functional magnetic resonance imaging, fMRI.)

The United States was beaten in the hunt for the weak bosons, and now it would be beaten in the hunt for the Higgs. Although Fermilab made a valiant effort with its Tevatron collider, the focus shifted to Europe, first with the Large Electron-Positron Collider (LEP), and then with the Large Hadron Collider at the CERN laboratory in Geneva. Perhaps American physicists might have been smarter to follow the CERN model, where many countries play major roles and contribute money. Why does the United States have to lead in everything, anyway? In any case, American physicists have participated significantly in the latest CERN experiments.

After a huge effort, LEP failed to find the Higgs and was shut down in 2000 so its tunnel could be used for the LHC. The Tevatron was terminated in 2011. While the LHC is huge and still costs billions, it is smaller and cheaper than the SSC would have been, using an already-existing tunnel "only" 27 kilometers long compared to 87.1 kilometers and "only" 14 TeV total collision energy compared to 40 TeV for the SSC. In his book, Lederman had predicted that the Higgs would be seen at the SSC by 2005. It certainly would have settled the matter one way or the other and would have probed far deeper than the LHC will be able to.

This is to not to take any credit away from the LHC and the thousands of dedicated and talented scientists and engineers who have brought it into being. The LHC first began circulating beams of protons in opposite directions on September 10, 2008. It is currently operating at 4 TeV per proton beam and is scheduled to run at 7 TeV per beam in 2014, after a shutdown of twenty months for the upgrade.

Of the six detectors assembled at the LHC, two—ATLAS (A Toroidal LHC Apparatus) and CMS (Compact Muon Solenoid)—are general-purpose detectors that are involved in the hunt for the Higgs and other new phenomena. As of the Higgs announcement, the ATLAS collaboration consisted of 3,000 scientific authors from 174 institutions in 38 countries. CMS consisted of 3,275 physicists (of which 1,535 are students), and 790 engineers and technicians, from 179 institutes in 41 counties.

HIGGS CONFIRMED!

On July 4, 2012, CERN announced that both ATLAS and CMS have seen a signal in the mass range of 125 to 126 GeV that is very likely the long-sought-after Higgs boson, or a reasonable facsimile. The statistical significance in each case was reported as "5-sigma," which implies that the probability it was a statistical fluctuation is one in 3.5 million. That is, each experiment taken alone was highly significant statistically. Even more important was the independent replication. Both the two-photon and four-lepton decay modes were seen.

It still remains to be demonstrated conclusively that this is indeed the Higgs boson and not some composite object made of already-known particles. So far it looks as expected for the Higgs of the standard model, but that may change as more data come in.

In late July 2012, ATLAS[20] and CMS[21] submitted papers to *Physics Letters* that provided more details and improved statistical significance.

MASS

I need to correct a common misunderstanding about the role that the Higgs boson, if indeed it exists, plays in the universe. Media reports, and even some public statements by physicists, have left the impression that the Higgs is responsible for the mass of the universe. This is not true. It is actually responsible for only a small fraction of the total mass of the universe.

This is not to say that the Higgs boson is not important. The main role of the Higgs in the standard model of elementary particles is to provide for the symmetry breaking of the unified electroweak force by giving mass to the weak bosons and splitting the electromagnetic and weak nuclear forces. It also gives mass to the other elementary particles, except for the massless photon and gluon. If elementary particles did not have mass, they would all be moving at the speed of light and would never stick together to form stuff like stars, cats, and you and me.

The mass of the universe is not simply the sum of the masses of the elementary particles that constitute matter. As we have seen, Einstein showed that the mass of a body is equal to its rest energy. If that body is not elementary but composed of parts, then its rest energy as a whole will be the sum of the energies of its parts. This sum will include the kinetic and potential energies of the parts in addition to their individual rest energies.

Now, for the bodies of normal experience, such as your neighbor's cat, the kinetic and potential energies of their parts are small compared to their rest energies. So, for all practical purposes, the total mass of a cat is equal to the sum of the masses of its parts.

This is even true at the microscopic scale. The masses of chemical elements are, typically, thousands of MeV, while the kinetic and binding energies are a few tens of eV. Only when you get down to the nuclei of the chemical elements do you get kinetic and potential energies that are measurable fractions of rest energies.

Inside nuclei, we have nucleons—protons and neutrons—that are themselves composed of quarks. Since quarks do not appear as free particles outside nucleons, we have to estimate their masses from studying the effects of their mutual interactions on the properties of nucleons and other particles that are composed of quarks. Fortunately, there are only six quarks but hundreds of particles (hadrons) made from these quarks to provide data to pin down quark properties. By using supercomputers to make numerical calculations referred to as *Lattice QCD*,[22] physicists have obtained reliable estimates of quark masses, which are given in table 11.1. The result: the masses of the quarks inside a proton or neutron constitute only 1 percent of its mass. The other 99 percent is due to the potential energy of the strong force.

The objects familiar to most humans, including most scientists, have masses that are essentially given by the number of protons and neutrons they contain. So we can say that only 1 percent of that mass arises from the masses of quarks. Furthermore, this normal stuff is itself only 5 percent of the total mass of the universe. So the Higgs contribution to the mass of the universe, unless it has something to do with dark matter or dark energy (see chapter 12), is at most one part in two thousand.

GRAND UNIFICATION

In the decades since the development of the standard model, numerous attempts have been made to figure out what might come next as we explore to ever higher energies and smaller distances. The first step was to build on the successes of the past. When we look at the history of physics, one grand principle especially stands out, that is, *unification*.

When Newton realized that the same force was responsible for an apple falling to the ground as was responsible for the moon falling around Earth, he unified terrestrial and celestial gravity. When Faraday and Maxwell showed that electricity and magnetism were the same phenomenon, they unified electromagnetism. When Salam, Glashow, and Weinberg developed the electroweak model, they unified the electromagnetic and weak nuclear forces.[23]

In the standard model, the strong nuclear force exists independently of the electroweak force, yet both are gauge theories. So it was obvious to attempt to unify these forces in what became known as GUTs for *Grand Unified Theories*.[24] As we have seen, symmetry groups play an important role in the standard model. In a unification scheme, some higher-order group, that is, one with many degrees of freedom, is assumed to describe the underlying symmetry, which then breaks down to give the structure we observe at lower energies. This is precisely the situation we described for the electroweak force. At high energies, the electromagnetic and weak forces are unified and their gauge bosons are all massless. The symmetry group is $SU(2) \times U(1)$. At lower energies, the electroweak force breaks down into electromagnetism, whose symmetry group is U(1), and the weak force, whose symmetry group is SU(2).

The problem with grand unification is that there are many groups to choose from, and without data, you can only speculate. In 1974, Sheldon Glashow and Howard Georgi made the reasonable proposal to start with the simplest symmetry group that could break down to the standard model, what is called *minimal* SU(5). Of all the other proposed GUTs, only minimal SU(5) made predictions that were testable with existing technology. But this prediction was an important one. Most theories that push beyond the standard model anticipate that the proton must be ultimately unstable. Otherwise, we cannot account for the fact that there exists a billion times as much matter as antimatter in the universe. The standard model makes no distinction between the two.

Minimal SU(5) made the specific prediction that the proton would decay into a positron and neutral pion,

$$p \rightarrow e^+ + \pi^0,$$

at a rate of one every $10^{30\pm1}$ years. Many other decay modes are possible, but this one has the shortest lifetime as well as detectable decay products. The π^0 decays almost immediately into two gamma-ray photons.

Now, we can't watch a single proton for $10^{30\pm1}$ years, but ten tons of water contain 6×10^{30} protons, so if we watch this much water for a year, we might see six protons decay. In the early 1980s, several experiments were mounted to search for proton decay. They were all installed deep underground, where the background from cosmic-ray muons was greatly reduced. The most sensitive of these, located deep in mines in Cleveland, Ohio, and Kamioka, Japan, contained 3,300 tons and 1,040 tons of purified water, respectively. However, none of these experiments, or any of several others I have not mentioned, witnessed proton decay. Thus, they effectively falsified minimal SU(5).

Incidentally, some philosophers of science argue that theories in physics are never falsified because their authors always fiddle things around to make them still consistent with the data. That claim is falsified by this example. While other GUTs were not ruled out, minimal SU(5) definitely was. There was no fiddling to be done. Note that a negative result can often be just as important as a positive one, as was the case here. These experiments were far from failures.

In 1996, the Kamioka detector was expanded to 50,000 tons of purified water and dubbed "Super-Kamiokande" (Kamiokande = Kamioka Nucleon Decay Experiment). I mentioned this experiment, in which I collaborated, earlier in this chapter where I discussed how it found evidence for neutrino mass in 1998. Super-K is still in operation and continues to search for proton decay. I left the project in 1999 shortly before retiring from the University of Hawaii. Results published in 2009 place a lower limit on the proton lifetime for the above reaction of 8.2×10^{33} years.[25] A number of other decay channels have been explored with similar negative but still extremely useful results.

SUPERSYMMETRY

As we have seen, all particles, elementary or not, divide up into two categories: *bosons*, such as the photon, Higgs boson, and helium atom, that have zero or integral spin; and *fermions*, such as the electron and lithium nucleus, that have half-integral spin. The principle of *supersymmetry* (SUSY) proposes that the basic laws of physics do not distinguish between bosons and fermions.

SUSY therefore predicts that every known boson has a "spartner" fermion and every known fermion has a "spartner" boson. The fermion electron is spartnered with a boson *selectron*. The boson photon is spartnered with a fermion *photino*. The fermion quark is spartnered with a boson *squark*. You get the idea. (Don't physicists have fun with the names they give things?) Supersymmetry implies that we can take, for example, the equations of QED that allow us to calculate the interaction of photons and electrons to high precision and apply them to the interaction of photinos and electrons, photons and selectrons, and photinos and selectrons.

Now it is clear that, like electroweak symmetry, supersymmetry is broken in our current universe; otherwise, the spartners of known particles would be just as common as the particles themselves because they would have the same masses, low enough to be detected long ago. A selectron would have the same mass as an electron. Since no one has detected any of these particles yet, they must be very massive (or nonexistent).

Indeed, theoretical estimates indicate that SUSY should show up at the LHC. As of this writing, when the collider has already been in operation for two years and has already seen what looks like the Higgs boson, no sign of SUSY has been found. In fact, when this book was in press, unpublished results were trickling in that, for all practical purposes, suggest SUSY is being ruled out by the failure to detect any sparticles in the expected mass range. If the LHC fails to detect SUSY, much of the theoretical effort over the past two or three decades will have been largely wasted. In particular, string theory will have to be abandoned as the ultimate "theory of everything" (TOE). I wouldn't be disappointed, though, because it would mean that we are finally learning something new.

12

ATOMS AND THE COSMOS

If God created the world, where was he before the Creation? . . . Know that the world is uncreated, as time itself is, without beginning and end.
                        —Mahaprana (India, ninth century)

AFTER THE BANG

Atomism postulates a universe composed of submicroscopic particles that possess the measurable quantities of mass, energy, and momentum that characterize the inertia we associate with matter. They kick back when you kick them. So far, we have focused on the nature of the particles themselves and the history of how our current understanding of their nature has come about. We have seen how the reductionism explicit in the atomic model was, and still is, widely unpopular with those whose philosophical and theological sentiments embrace a more holistic picture of a universe in which everything is closely connected to, and dependent on, everything else along with, perhaps, some outside power.

Atoms, by contrast, are only weakly connected to one another. They move around mostly independently in mostly empty space and interact only when they collide with one another, occasionally sticking together to form composite masses. Because we and the stuff around us are examples of such composite masses, we tend to think this is an important characteristic of matter. In fact, the universe is largely composed of particles in random motion that rarely interact. And there is no holism or outside power in atomism.

Only when the evidence became unassailable did the antiatomists finally admit, at least, that atoms serve as a good model for describing scientific observations and may even be "real." Even now, however, the ancillary implications of atomism explicit in the philosophy of Democritus, Epicurus, and Lucretius are not palatable to many of those who otherwise accept the atomic model. Atomism, after all, is atheism.

The early atomists imagined a cosmos that was quite different from the conventional view that grew out of religious tradition. Most cultures have assumed a universe finite in size that came into existence at a finite time in the past by means of the purposeful act of a Creator, or perhaps some impersonal but still supernatural force. In the common theological view, not only matter and energy but also space and time themselves came into being at that pregnant moment.

Most who read this book will be familiar with the creation story in Genesis from the Hebrew Bible and Christian Old Testament. The Qur'an tells a similar tale. In 1927, a Belgian Catholic priest, Georges Lemaître (1894–1966), first proposed what was later dubbed the "big bang."[1] That is, he showed that Einstein's equations of general relativity implied an expanding universe that resembled an explosion.

Popes, theologians, and other religious figures seized on the big bang as providing scientific verification of a divine creation. In 1951, Pope Pius XII told the Pontifical Academy, "Creation took place in time, therefore there is a Creator, therefore God exists."[2] Lemaître, a good scientist as well as a priest, wisely advised the pope not make this statement "infallible."

Even Einstein did not initially believe Lemaître, despite the fact that Lemaître had used Einstein's own equations to infer that the universe was expanding. Einstein still held to the almost-universal

belief, found in Genesis and elsewhere, that the universe is composed of a "firmament" of more or less fixed stars that remain, on
average, the same distance apart. His gravitational equation has
a constant term in it that, when positive, results in a gravitational
repulsion. He called this the *cosmological constant* and conjectured
that it provided the repulsive gravitational force needed to balance
the familiar gravitational attraction and keep the universe stable.

Although this is not widely known, Lemaître had data to back
up his proposal. Unfortunately, his paper was in French and was
not immediately translated into English.[3] Independent confirmation, in English, came in 1929 when Edwin Hubble (1889–1953) in
California determined from his and other's observations that the
galaxies are, on average, moving away from us at speeds roughly
proportional to their distances. That is, when you make a plot of the
recessional speeds of galaxies versus their distances, what is called
the *Hubble plot*, the data points scatter around a straight line. This
was just what was to be expected if the galaxies were remnants of a
giant explosion that occurred in the past, now estimated to be 13.75
billion years ago.[4]

Once the universe's expansion was confirmed, it became generally assumed that the cosmological constant is identically zero.
Einstein called his inclusion of the constant in his theory his "greatest
blunder."[5] However, it was certainly not the "fudge factor" that it is
usually portrayed as in the literature. This is a derisive term applied
to an arbitrary quantity added to an equation to make it agree with
the data. The cosmological constant was already in Einstein's equation, and he had no theoretical reason to set it to zero.

INFLATION

When Einstein's gravitational equation, including a positive
cosmological constant, is applied to a universe empty of matter,
it predicts an exponentially expanding universe. Here, the term

matter includes familiar objects with mass as well as electromagnetic radiation, which we have seen is a form of matter (photons). In the 1980s, several physicists independently proposed that during a tiny interval of time (10^{-35} second or so) after the universe first appeared, it underwent a rapid, exponential expansion called *inflation*.[6]

The expansion observed in the data by Lemaître and Hubble was far from exponential. Rather it appeared to be linear. That is, when the recessional speeds of galaxies were plotted against their distances, they appeared to be scattered around a straight line. This is exactly what you would expect in an explosion when the pieces go flying off independently of one another. The faster pieces will go farther.

This picture, however, was inconsistent with other observable facts. First, the universe appears very homogeneous and isotropic, as if it has reached some kind of equilibrium. But the farthest galaxies on opposite sides of the universe traveling near the speed of light relative to one another could never have been in causal contact in the linear scenario. That is, in the linear scenario, the universe did not have time to equilibrate. This was called the *horizon problem*.

Second, the measured average density of matter in the universe seems to be precisely equal to the critical density at which the universe is geometrically flat. That is, although general relativity describes space-time in terms of non-Euclidean geometries, the data indicate, on average, a Euclidean universe. This was called the *flatness problem* because Euclidean space, such as a plane, is flat, while non-Euclidean space, such as the surface of a sphere, is curved. A flat universe is very unlikely in the linear scenario.

Furthermore, Grand Unification Theories (GUTs) implied that the universe is filled with magnetic monopoles (see chapter 6) produced in the early universe. Yet, after intense searches, none have been found. This was called the *monopole problem*.

Inflation solved all three problems. During the inflationary phase, the universe expanded so enormously that the galaxies in all directions and at all distances came from the same original region of space. This solved the horizon problem. To see how the flatness problem was solved, think of the surface of a balloon blown up to huge size. A small patch on the surface of the balloon will be very flat. Our universe is like that patch. Note that other patches exist on the rest of the surface. The full surface of the balloon represents the universe that, in the inflationary scenario, must exist outside our *light horizon*. The light horizon defines the distance within which light can have reached us in 13.7 billion years. One estimate is that 10^{100} as many galaxies exist there, beyond our reach, as the 100 billion or so within our horizon.

As for magnetic monopoles, because they were produced only in the early universe, inflation spread them over the entire universe, including the enormous part now beyond our horizon. The chance of finding one is negligible. That doesn't mean we should stop looking. If some are found, inflation might be in trouble.

Inflation also solves another problem. It provides a mechanism for the formation of galaxies. Observations by a series of increasingly more sophisticated space-born instruments, notably the Cosmic Background Explorer (COBE)[7] and the Wilkinson Microwave Anisotropy Probe (WMAP),[8] have provided incredibly precise measurements of the cosmic background radiation. This is a thermalized gas of the very low-energy photons left over from the big bang that fills the universe. These photons have been cooled to 2.7 degrees Kelvin by the universe's expansion and are in the microwave region of the electromagnetic spectrum.

While highly uniform and isotropic, observations reveal a small fluctuation of one part in one hundred thousand. Cosmological models that combine inflation and the (almost) linear big bang nicely account for just the right amount of matter clumping to form the first galaxies around 13 billion years ago.

I don't want to sweep under the rug problems with the inflationary model, which has a number of prominent opponents. However, their objections are theoretical and philosophical rather than observational. No noninflationary model exists that answers as many questions and fits all the data as well as inflation.

Let me list some of the achievements of WMAP, which quantitatively define the universe far better than ever before:[9]

- It resolved the full microwave sky to 0.2 degree.
- It determined the age of the universe to be 13.75 ± 0.11 billion years.
- It showed that the universe is Euclidean to within 0.6 percent.
- It showed that familiar atomic matter makes up only 4.6 ± 0.2 percent of the universe.
- It showed that 22.7 ± 1.4 percent of the mass of the universe is *dark matter* not made of atoms.
- It determined that the *dark energy* makes up 72.8 ± 1.6 percent of the universe.
- It measured the polarization of the cosmic background radiation and provided details on the nature of the fluctuations that formed the galaxies.
- It ruled out several models that attempted to describe the universe when it was less than one-trillionth of a second old.
- It greatly improved the limits on the adjustable parameters of cosmological models.

THE STUFF OF THE UNIVERSE

In this book, we have worked our way from the speculations of the ancient Greek philosophers on the nature of matter to our current understanding in terms of the elementary particles of the standard model. We saw that physicists who work at particle colliders deal with hundreds of varieties of tiny objects that they describe with three generations of quarks, leptons, and gauge bosons (see table 11.1). For most scientists, the situation is even simpler; everything they deal with can be accounted for by four particles: the u and d quarks, the electron, and the photon. They deal with only matter on Earth. Astronomers, cosmologists, and particle physicists have to deal with the matter in a whole universe. What do they find?

Let us begin with the matter that composes the stars and planets. That turns out to be the same simple stuff, chemical elements made of the same quarks and electrons, as we see on Earth. Not terribly interesting to the cosmologist, but that's not all there is. In fact, the luminous matter in stars comprises only 0.5 percent of the mass of the universe.

In the 1930s, astronomers noticed that the orbital speeds of stars at the edge of the Milky Way do not have the values that they should have according to Newton's law of gravity and the observed distribution of visible matter in the galaxy. In fact, their orbits are best described as if they are moving through invisible matter of more or less uniform density and extending well beyond the luminous stars in our galaxy. In the years since, other observations in the Milky Way and beyond have indicated that about 23 percent of the mass of the universe is comprised of this invisible form of matter dubbed *dark matter*.

Besides the measurements of stellar motions, a number of other types of observations have confirmed the existence of dark matter.[10] One of the most convincing and exotic is *gravitational lensing*. Galaxy clusters are dense groups of hundreds of galaxies held together by gravity. Light from objects behind a cluster can be bent by this strong gravity. A lensing effect sometimes occurs, resulting in multiple images of the source in our telescopes. Measurements on such images enable the mass of the clusters to be estimated. These turn out to be much higher than the mass estimated from the emitted light.

As we saw above, WMAP has also confirmed the existence of dark matter and has provided the best estimate of its contribution to the mass of the universe.

WHAT IS THE DARK MATTER?

Let us ask, what should be the properties of whatever entities comprise dark matter? First of all, they must interact very weakly with normal matter, or else they would have been detected by now. For the same reason, they must be electrically neutral. Second, they are likely to be massive in order to produce their large gravitational effects, although it's possible that they are very light and there are just many of them.

A variety of cosmological evidence implies that dark matter cannot be composed of the same elementary particles that have been identified as the ingredients of all the matter we observe on Earth, including the most exotic particles that have so far been produced in colliding-beam experiments or observed in cosmic rays. This matter is referred to as *baryonic* because most of its mass is in the heavier particles such as the proton and neutron, which are classified as baryons.

The evidence for the dark matter not being baryonic requires some explanation. It is important to one of the themes of this book, which is to demonstrate the great extent of our current knowledge of the nature of the matter in our universe.

We have covered in great detail the highly successful standard model of elementary particles and forces. In the last few decades, data from Earth- and space-based telescopes covering the full electromagnetic spectrum, from radio to gamma rays, have enabled cosmologists to develop a standard model (with variations) of their own to describe this wealth of data.

The standard model of cosmology includes a quantitative description of the synthesis of the nuclei of the light chemical elements hydrogen, helium, and lithium in the early big bang. The two isotopes of hydrogen, $_1H^1$ and $_1H^2$, and of helium, $_2He^3$ and $_2He^4$, are included (see chapter 8 for notation). The abundances of these nuclei are very sensitive to the baryon density of the universe, which is estimated to be about 5 percent of the observed density of all matter, including 4.5 percent nonluminous baryonic matter. Thus, 95 percent of the matter of the universe is nonbaryonic. As mentioned, the dark matter is about 23 percent while the rest is dark energy, which will be covered in the next section.

The favorite candidates for the dark matter are WIMPs, *weakly interacting massive particles*.[11] Several theories for new physics at energies beyond about 1 TeV, where electroweak symmetry is no longer broken, in particular supersymmetry (see chapter 11), predict a new, stable WIMP at a mass of about 100 GeV. The favorite SUSY candidate is the *neutralino*, a combination of the spartners of the photon, Z boson, and neutral Higgs boson. SUSY also predicts several Higgs bosons, including some that are charged, not to mention the *higgsino*, the spartner of the Higgs.

Searches for dark matter are underway in both collider and deep underground laboratories.[12] It can be anticipated that the LHC will either confirm supersymmetry or dispose of it. Since it has already produced a 125 GeV Higgs, a WIMP of about the same mass is well within its capability. Physicists were surprised and delighted that the Higgs candidate appeared in only two years of running. They expected that SUSY would be seen first, and this did not happen. In fact, as mentioned in the previous chapter, doubts are beginning to surface on the validity of SUSY. Hopefully the issue will be settled before the LHC shutdown in 2013. If not, it certainly will be settled after the shutdown, since the machine energy will be doubled.

Other non-SUSY candidates for the dark matter include *sterile neutrinos*. Sterile neutrinos could constitute both the ingredient of dark matter and a pointer to new physics.[13] As mentioned earlier, it was demonstrated in 1998 that neutrinos have mass. Massless particles with spin have the feature of always spinning either in the same direction as their motion, like a right-handed screw, or opposite. Neutrinos are observed to have left-handed "helicity," that is, to spin opposite to their direction. Antineutrinos are right-handed, like a normal screw.

However, when a particle has mass, it travels at less than the speed of light, and you can always find a reference frame with either helicity. It follows that neutrinos must have a small right-handed component while antineutrinos have a small left-handed part. Since these components are not observed, they may possibly be "sterile," meaning, they interact mainly gravitationally. In that case, they are good candidates for dark matter.

Interest in sterile neutrinos has been piqued by several hints in neutrino experiments and astronomical observations.[14] None are sufficiently significant, however, to claim a discovery.

Sterile neutrinos would have to have a mass of a few keV (1 keV = 1,000 eV), much heavier than normal neutrinos but still much lighter than electrons, to constitute the dark matter. The main decay mode of a sterile neutrino is into two normal neutrinos and an antineutrino, which is not very detectable, to say the least. However, the decay into one normal neutrino and an x-ray photon is detectable in the diffuse cosmic x-ray background, at least in principle. To constitute the dark matter, the average lifetime of the sterile neutrino must be more than a billion years.[15]

Another long-considered candidate is a hypothetical particle called the *axion*. This particle would help solve some technical problems with quantum chromodynamics. Laboratory measurements constrain its mass to about 10^{-4} eV.[16]

DARK ENERGY

We have seen that the recessional speeds of galaxies have indicated a simple linear relationship between distance and speed in the Hubble plot. While speeds are easy to measure from the redshifts of spectral lines, measurements of distances are much more difficult and, until recently, were very crude. In the late 1990s, two independent projects were able to improve the accuracy of distance measurements to galaxies. Using ground-based telescopes, astronomers studied exploding Type 1a supernovas inside galaxies.

Such supernovas provide excellent "standard candles" because of the consistency of their peak luminosities. Type 1a supernovas result from the explosion of white dwarf stars, which occur only in a narrow range of stellar mass.

Both groups expected to find evidence for the plot of distance versus speed to start turning down at great distances as gravity acts to slow the expansion. Instead, they got the surprise of their lives (and the Nobel Prize to boot): the curve turned up instead of down.[17] They found that, using the Type 1a supernovae as standard candles, galaxies at a given redshift (higher recessional speed) were farther away than expected, indicating that the universe is undergoing an acceleration of its expansion rate. The universe is falling up![18]

Since then, this result has been amply confirmed by the Hubble Space Telescope and other observations. The expansion of the universe is speeding up. Furthermore, whatever is responsible for this acceleration is the main ingredient of the universe, its energy density constituting 73 percent of the mass/energy density of the universe. This ingredient is called *dark energy* to distinguish it from dark matter, although both are fundamentally material in nature.

While surprising, the result was not totally unanticipated or completely mysterious. As we saw above, Einstein's cosmological constant, if positive, will produce a gravitational repulsion.[19] Further observations so far are consistent with the interpretation that the cosmological constant, which is equivalent to an energy density, is the source of dark energy. However, other possibilities for the source of the cosmic acceleration[20] are actively under consideration because, as we will see next, the cosmological constant option has a serious difficulty.

Alternatively, a gas can have negative pressure and gravitational repulsion under the right conditions. A field of particles might exist throughout the universe, somewhat like the Higgs field, that has negative pressure and produces gravitational repulsion and the acceleration of the expansion of the universe.

THE COSMOLOGICAL CONSTANT PROBLEM

In 1989, Steven Weinberg published a paper showing that a huge discrepancy existed between the observed upper limit on the cosmological constant and the value calculated from elementary particle physics.[21] The cosmological constant is equivalent to a field of constant energy density filling the vacuum of space. Recall from chapter 9 that quantum fields have a zero-point energy that corresponds to a state in which no quanta of the field are present. This can be interpreted as the energy density of the vacuum.

The calculation of that energy density is very straightforward.[22] Assuming only photons or other bosons, the result is 120 orders of magnitude higher than what is observed. This is the number you usually see in the literature. However, recall that fermions have a negative zero-point energy and when they are included, the discrepancy is lower. Nevertheless, the difference is still at least 51 orders of magnitude, so the problem remains a problem. This calculation is often characterized as the worst in physics history!

Weinberg and others have considered various ways out. However, we can think of at least one reason why the calculation has to be wrong, other than the obvious fact that it is so grossly inconsistent with the data. The energy-density estimate is made by summing up all the quantum states throughout a sphere within the universe. However, the number of states in such a sphere cannot be greater than that of a black hole of the same size. And it can be shown that the number of states of a black hole is proportional to its surface area, not its volume. When the calculation is redone summing over the states on the surface of the sphere, the resulting energy density is very close to the density of the universe.[23] Because the density of the dark energy is 73 percent of the total, this is good agreement indeed.

However, I may have oversimplified the situation, since this solution to the cosmological constant problem is far from achieving a consensus. In any case, a number of other solutions have been proposed, and we just have to await one that satisfies the experts.

BEFORE THE BANG

The theologians who waxed so enthusiastic about the big bang were well aware that the cosmological model bore little, if any, resemblance to the creation story in Genesis or that of any other religious tradition for that matter. The key point for them was that science was providing evidence that a beginning did in fact occur, that there was a moment in the past when the universe appeared out of nothing: *creation ex nihilo*.[24]

Furthermore, as shown in 1970 by Stephen Hawking and Roger Penrose, Einstein's general theory of relativity implies that the universe at its first moment of existence was a *singularity*, that is, an infinitesimal point in space of infinite energy density.[25] This meant that not only was matter created at that moment, but so were space and time. As we saw in chapter 2, Augustine addressed the question of what God was doing in all that time before he created the universe by saying that God created time along with the universe.

A finite, created universe conflicts with the teachings of the atomists that the universe is eternal and boundless. In chapter 1, Epicurus was quoted as saying, "The universe is without limit." The big bang seemed to refute atheist atomism.

However, there was a fly in the ointment. General relativity is not a quantum theory and so does not apply to a region of space less than 1.616×10^{-35} meter in diameter, called the *Planck length*, named for the physicist Max Planck who, as we saw in chapter 7, initiated the quantum revolution. Applying the Heisenberg uncertainty principle of quantum mechanics, it can be shown that it is fundamentally impossible to define a smaller distance or to make any measurements inside a region of that size.[26] Basically, we can have no information about what is inside a sphere with a diameter equal to the Planck length. It is a region of maximal chaos.

The uncertainty principle also mandates that no time interval shorter than 5.391×10^{-43} second, called the *Planck time*, can be measured. Thus, our cosmological equations, derived from general relativity, can apply only for times greater than the Planck time and only for distances greater than the Planck length. Although their singularity proof was correct for the assumptions made, both Hawking and Penrose long ago agreed that it does not apply once quantum mechanics was taken into account,[27] a fact most theologians, including William Lane Craig, have conveniently ignored. In short, the origin of our universe was not a singularity and need not have been the beginning of time.

Think of a time axis as extending without limit from time $t = 0$. If you use the operational definition of time as what you read on a clock, rather than giving it some unmeasurable metaphysical meaning, the minimum definable time interval is the Planck time. Thus, the time parameter, t, will not be a continuous variable but will be discrete in steps equal to the Planck time. That is, according to quantum mechanics, time is quantized in units equal to the Planck time: $t = 0, 1, 2, 3, \ldots$ and so on.

Because the Planck time is so small by ordinary standards, we can generally get away with treating t as a continuous variable—even in most quantum mechanical applications. However, we cannot assume temporal continuity as we move closer to the origin of our universe. In fact, to properly describe what happens, we need to use a theory in which quantum mechanics and general relativity are combined. Despite decades of effort, no such theory, usually referred to as *quantum gravity*, has yet to be successfully formulated.[28] I am not sure what such a theory will accomplish anyway, since we already have established that we cannot measure anything at distance intervals less than the Planck length and at time intervals less than the Planck time.

In any case, if we stick to intervals greater than the Planck time, we are probably on safe ground to at least draw some general conclusions based on our best existing knowledge. Note that we can also count time steps in the opposite direction: $t = -1, -2, -3, \ldots$ and so on. Nothing in the cosmological equations derived from general relativity prevent them from being applied for $t < 0$. That is, we cannot rule out another universe on the opposite side of our time axis.

In one scenario, which I have discussed in previous books and has been worked out mathematically, our universe appears from an earlier one by a process known as *quantum tunneling*.[29] For our purposes here, suffice it to say that nothing in our current knowledge of physics and cosmology requires us to conclude that the beginning of our universe was the beginning of space, time, and everything else that is.

THE MATTER-ANTIMATTER PUZZLE

Besides the nature of dark matter and dark energy, another important puzzle about the universe that remains to be solved is its large excess of matter over antimatter. If the universe began, as cosmology and physics suggest, in perfect symmetry, then why is there such a large asymmetry between matter and antimatter? For every billion protons and electrons found in the universe, there are only one antiproton and one antielectron (positron), along with a few produced in high-energy cosmic-ray collisions.

In 1967, the great Russian physicist and political dissident Andrei Sakharov suggested a solution.[30] He proposed that baryon-number conservation and CP symmetry are violated, leading to matter-antimatter asymmetry. Recall from chapter 10 that the violation of CP symmetry was observed in 1964. In chapter 11 we saw that the decay of protons, which violates baryon-number conservation, is predicted by some Grand Unified Theories. Despite intense searches, proton decay and the violation of baryon-number conservation so far have not been observed.

Nevertheless, we still have a reasonable expectation that the violation of baryon-number and lepton-number conservation will eventually be discovered and we will be able to explain the matter-antimatter asymmetry in the universe.

THE ETERNAL MULTIVERSE

As we saw previously, while the big bang was the beginning of *our* universe, it was not necessarily the beginning of all that is. Modern inflationary cosmology strongly suggests that other universes besides our own exist in what is called the *multiverse*.[31] Because we have no observational evidence (yet) for other universes, I will not indulge in speculations about them, except to say that such speculations are based on well-established science and their ultimate empirical confirmation is not out of the realm of possibility. In any case, allow me to simply use the term *multiverse* to refer to all that is, even of it should turn out that our universe is all there is.

Still, I must address a question that more closely relates to the cosmological view of the atomists, which is whether a multiverse (all that is) can be eternal. While the ancient atomists accepted the fact that everything made of atoms eventually decays, the atoms themselves were eternal and continued to make worlds anew. Although I have covered this question in earlier books,[32] I need to review it here because of its direct relevance to atomism. I can do this in a few words.

William Lane Craig and other theistic philosophers and theologians have argued that an eternal multiverse is mathematically impossible.[33] Simply put, they claim that if the multiverse is infinitely old, then it would have taken an infinite time to reach the present. That is, we would never reach "now."

My rebuttal is equally simple. An eternal multiverse did not begin an infinite time ago. It had no beginning. It always was. No matter how far you go back in time—100 billion years, 100 trillion years—the time to the present is finite. An eternal multiverse is perfectly possible, mathematically and scientifically.

SOMETHING ABOUT NOTHING

I am often confronted with a rhetorical question from theists that they are convinced settles the argument about the existence of a supernatural creation: "How can something come from nothing?" Well-known cosmologist and popular author Larry Krauss attempted to provide an answer in a 2012 book, *A Universe from Nothing: Why There Is Something Rather Than Nothing*.[34]

Krauss describes how our universe could have arisen naturally from a preexisting, structureless void he calls "nothing." He bases his argument on quantum physics, along with now well-established results from elementary particle physics and cosmology. In an afterword, prominent atheist Richard Dawkins exults, "Even the last remaining trump card of the theologian, 'Why is there something rather than nothing?' shrivels up before your eyes as you read these pages."

Philosopher and physicist David Albert is unimpressed. In a review in the *New York Times* he asks, "Where, for starters, are the laws of quantum mechanics themselves supposed to have come from?"[35] Krauss admits he does not know, but he suggests they may arise randomly, in which case some universe like ours would have arisen without cause. In my 2006 book *The Comprehensible Cosmos*, I show how the primary laws of physics, notably the great conservation principles, arise naturally from the symmetries of the void and the spontaneous breaking of those symmetries.[36] We have already seen how symmetries provide the foundation of the standard model of particles and fields.

In any case, Albert asserts that it doesn't matter what the laws of physics are. He insists they "have no bearing whatsoever on questions of where the elementary stuff came from, or of why the world should have consisted of the particular elementary stuff it does, as opposed to something else, or to nothing at all."[37]

Albert is not satisfied that Krauss has answered the fundamental question: Why is there something rather than nothing, or, put differently, why is there being rather than nonbeing? Again, there is a simple retort: Why should nothing, no matter how defined, be the default state of existence rather than something? The theist has the burden of showing how there could have been nothing. And one could ask the theist: Why is there God rather than nothing? Once they assert that there is a God (as opposed to nothing), they can't turn around and ask a cosmologist why there is a universe (as opposed to nothing). They claim God is a necessary entity. But then, why can't a godless multiverse be a necessary entity?

Krauss says that the reason there is something rather than nothing is that the quantum vacuum state is unstable. Here he refers to the statement made by Nobel-laureate physicist Frank Wilczek in a 1980 article in *Scientific American*, "Nothing is unstable."[38]

Krauss's theological and philosophical critics claim that what he discusses is not really "nothing." Krauss dismisses this criticism and says that the "nothing" of his critics is some "vague and ill-defined" and "intellectually bankrupt" notion of "nonbeing." Albert retorts, "Krauss is dead wrong and his religious and philosophical critics are absolutely right."

Clearly, no academic consensus exists on how to define "nothing." It may be impossible. To define "nothing" you have to give it some defining property. But, then, if it has a property, then it is not nothing!

Krauss shows that our universe could have arisen naturally without violating any known laws of physics. While this has been well known for a quarter century or more,[39] Krauss brings the discussion up to date.

The "nothing" that Krauss mainly talks about throughout his book is, in fact, precisely definable. It should perhaps be better termed as a "void," which is what you get when you apply quantum theory to space-time itself. It's about as nothing as nothing can be. This void can be described mathematically. It has an explicit wave function. This void is the quantum gravity equivalent of the quantum vacuum in quantum field theory.

So the real issue is not where our particular universe came from but where the multiverse came from. This question has an easy answer. As we saw above, the multiverse is most likely eternal. Repeating myself, since it always was, it didn't have to come from anything.

13

SUMMARY AND CONCLUSIONS

Religion comes from the period of human prehistory where nobody—not even the mighty Democritus who concluded that all matter was made from atoms—had the smallest idea of what was going on. It comes from the bawling and fearful infancy of our species, and is a babyish attempt to meet our inescapable demand for knowledge. Today the least educated of my children knows much more about the natural order than any of the founders of religion.
—**Christopher Hitchens**[1]

THEY HAD IT (MOSTLY) RIGHT

The view of nature proposed over two thousand years ago by the Greek philosophers Leucippus, Democritus, and Epicurus, and preserved for posterity by the Roman poet Lucretius, has been validated by modern physics, cosmology, and, where applicable, by the rest of science. However, I need to continually emphasize that what we are talking about here, whether from antiquity or from the twenty-first century, is a just a model that describes what we observe with our senses and scientific instruments. We have no way of knowing what is the ultimate reality that lies behind that model.

The atomic model, in which the universe is composed of matter and nothing else, is adequate to explain everything we observe and experience as human beings. In this model, matter itself is composed of elementary particles that move around in an otherwise completely empty void. Until the twentieth century, the particulate nature of matter was not directly observable but by the twentieth century was well confirmed empirically.

After its rediscovery in the Renaissance, atomism became the foundational principle of the scientific revolution of the seventeenth century. It reached fruition with the discovery in the nineteenth century that the chemical elements are well described as particulate, and in the twentieth century with the discovery that those elements are in fact composed of even more elementary constituents: quarks and electrons. At this writing, the latest triumph of atomism is the apparent confirmation in 2012 of the existence of the Higgs boson, which had been predicted forty-eight years earlier, as the source of the masses of elementary particles.

Because of its implicit and sometimes explicit atheism, the particulate model of matter has had many opponents throughout history—from Aristotle to Christian theologians, to nineteenth-century chemists and philosophers. While theists today must accept the fact that the evidence for the atomic theory of matter is overwhelming (just as is the evidence for evolution), they reject the theological implications of that theory—namely, atheist materialism.

Included in the assumptions of ancient atomism now confirmed by science are indeterminism and the dominant role of chance in the universe. Furthermore, our universe is very possibly just one of many in a multiverse unlimited in space and eternal in time. Any gods that may exist play no role in nature, including human life. The atomists also envisaged the evolution of life and a kind of survival of the fittest.

In short, they pretty much had it right, at least in general terms. The job of science since has been to fill in the details.

MATTER

In today's science, we find no evidence for any ingredient in nature other than matter. If some other immaterial substance exists, such as what is usually referred to as *spirit*, it has no effect on our senses or instruments that we can verify scientifically. Nowhere can you point to an observation or measurement that requires us to introduce some immaterial substance into our models in order to explain the data. Where we have observations for which no full explanation exists, such as the nature of dark matter and dark energy, these phenomena still produce measurable effects that identify them as material. We have measured the mass/energy density of these components. We know they exist by their gravitational effects. Presumably, gravity doesn't act on spiritual stuff.

Matter is defined as a substance with inertia. When you kick a chunk of matter, it kicks back. A material body is characterized by three measurable properties: mass m, energy E, and momentum $\mathbf{p}$. Momentum is a three-dimensional vector whose magnitude $|\mathbf{p}|$ we write as p and whose direction is the same as the velocity vector of the body. In units where the speed of light in a vacuum $c = 1$, $m^2 = E^2 - p^2$. While E and p depend on the reference frame in which their measurements are performed, m is invariant. It is the same in all reference frames.

Unfortunately, the term *energy* is much misused in common discourse and in mythology. An acupuncturist, trained in ancient Chinese medicine, claims to redirect the flow of "qi," an imagined form of energy, through your system. A touch therapist, often trained in an accredited nursing college, claims to adjust your "bio-energetic field." Today, the practice of complementary and alternative medicine is filled with unsupported claims of healing methods that use the term *energy* in one form or another, usually with the implication that it is associated with some spiritual force.[2]

However, as far as physics is concerned, it is incorrect to regard energy as something immaterial, something separate from mass. It is a property of matter just as is mass and momentum. In particular, light is fully material, composed (in our model) of particles called photons. Although a photon has zero mass, it has nonzero energy and momentum.

Our senses and instruments detect material objects and presumably are unable to sense spirit. However, this does not provide much credence to the hypothesis that immaterial objects might still exist. Of course, we can never fully rule out that they might show up someday. But in what sense can we say immaterial objects exist if they have no measurable effects on the material objects we do observe? And if they have no effect, who cares?

More important are the types of immaterial, supernatural entities that most people want to believe play a central role in nature and human life. In that case, they should produce observable effects on material objects. So far, after thousands of years of looking, we see no sign of such effects. This should rule them out beyond a reasonable doubt.[3]

The objects we call "atoms" today, the elements of the chemical periodic table, are now known to be composite and not truly elementary. Nevertheless, atomism achieved strong substantiation in the 1970s with the standard model of particles and forces that has agreed with all empirical data for four decades. In that model, familiar objects from cats to stars are composed of just three fundamental particles: u and d quarks that form the protons and neutrons in the nuclei of the chemical elements, and electrons that swarm in clouds around these nuclei. With the photon added to provide light and other electromagnetic radiation, we have everything that is involved in the life of almost every human on Earth (particle physicists and cosmologists excepted) reduced to the interaction of just four particles. If that isn't a triumph for atomism, I don't know what a triumph would entail.

In a high-energy collision, in cosmic rays or particle accelerators, many short-lived particles are produced as some of the energy of the collision is converted into mass. All these generated particles fit into a scheme that contains a total of six quarks and their partner anti-quarks and six leptons along with their antileptons. Additionally, twelve gauge bosons, including the photon, are responsible for the forces by which particles interact with one another: the electromagnetic force, the weak nuclear force, and the strong nuclear force. In the standard model, the electromagnetic and weak forces are united as a single electroweak force. It is split into the two distinct forces at the low energies of common experience, and, until recently, all laboratory experiments as well.

Gravity is currently treated separately from the standard model; it has negligible effect on the subatomic scale but ultimately must be included in any fully unified theory. However, all observed gravitational phenomena remain consistent with Einstein's 1916 general theory of relativity. In short, for decades now physics has had fundamental theories that agree with all observations. They just have not yet been combined into a single theory.

All the quarks, leptons, and gauge bosons of the standard model have been verified experimentally. Until July 4, 2012, the one gap remaining in the standard model was the Higgs boson, which had been proposed in 1964 to provide the mechanism by which elementary particles get mass. With the observation in 2012 of what looks very much like the Higgs of the standard model, that picture seems to have emerged victorious.

Nevertheless, the standard model is unlikely to be the final story. It requires over twenty parameters such as particle masses and force strengths that must be determined experimentally, as well as leaving many questions unanswered. That doesn't mean they ever will be. Perhaps the standard model is all we have and the rest is accident. In any case, it is hard to imagine ever reaching the time when every question has been answered. I have little confidence that a so-called theory of everything will ever be achieved, at least in the foreseeable future. But I may be wrong, which would be unfortunate because it would mean the end of physics.

None of these uncertainties detract from the continual success of the atomic picture. Even when they did not know anything about the chemical atoms, physicists succeeded in explaining the behavior of gases in terms of particles. Even when they did not know anything about the structure of these atoms, chemists built an enormously useful science based on the periodic table. Even when they did not know that the nucleons inside the nuclei of chemical atoms were composed of quarks, physicists developed nuclear energy. Each of these was an achievement for atomism.

MATERIALISM DECONSTRUCTED?

Now, those who read the popular literature might have received the impression that, in fact, modern physics has not confirmed the picture of atoms and the void or perhaps even refuted it. For example, in *The New Sciences of Religion: Exploring Spirituality from the Outside In and Bottom Up*, Christian apologist William Grassie says, "The concept of materialism deconstructed itself with the advent of quantum mechanics and particle physics."[4] To be ecumenical, he quotes the Hindu physicist Varadaraja V. Raman: "Physics has penetrated into the substratum of perceived reality and discovered a whole new realm of entities there, beyond the imagination of the most creative minds of the past."[5]

Now, maybe Democritus did not imagine quarks. But he imagined material particles, and quarks are material particles. The "new realm of entities" uncovered in modern physics is hardly beyond imagination. They are imagined in the quantum theory of fields, although just imagining something does not make it real—despite what some theologians claim and what some physicists seem to believe.

FIELD-PARTICLE UNITY

The claim that quantum mechanics has revealed a reality beyond matter is based on the mistaken notion that two separate realities exist: discrete, particulate matter and a plenum that is reminiscent of the long-discredited aether. However, the electromagnetic aether was at least material. The new aether is more abstract, more in tune with the duality of mind and body that is embedded in all religious thought. Unsurprisingly, theologians and spiritualists delight in this new dualism—handed to them on a platter by theoretical physicists.

You will often see it written that abstract, holistic quantum fields are the deeper reality while particles are simply the excitations of the fields. For example, in *The Atom in the History of Human Thought*, which I have referred to often, historian Bernard Pullman writes,

> To the extent that a Democritean influence has shaped our conception of the world, there has been a tendency to stress the corpuscular aspect of the standard model and to introduce a certain formal distinction between particles of matter and intermediary particles associated with force fields. As a result, we may have given the impression that this corpuscular aspect provides the most exact description of physical reality. Such a view would be unfortunate, as it might obscure what is considered today as the most plausible picture of reality, which not only unifies the concepts of particles and fields, but even considers fields preeminent over particles. . . . The fundamental and underlying reality of the world is embodied in the existence of a slew of fields and in their interactions."[6]

Unfortunately, many theoretical physicists have contributed to the impression that quantum mechanics has done away with the concept of matter. For example, in an article in *Scientific American*, physicist David Tong says:

> Physicists routinely teach that the building blocks of nature are discrete particles such as the electron or quark. That is a lie. The

building blocks of our theories are not particles but fields: continuous, fluid-like objects spread throughout space.[7]

Pullman and Tong are expressing the Platonic view of reality,
commonly held by many theoretical physicists and mathematicians.
In order to test their models, physicists assume that the elements
of these models correspond in some way to reality. But they are
compared against the data that flow from our so-called particle
detectors on the floor of an accelerator lab. It is the data that form the
concrete foundation of our knowledge. What is fundamental in our
model is not necessarily fundamental to our knowledge. Models are
squiggles on the whiteboards in the theory section of the physics
building. Those squiggles are easily erased; the data can't be.

Indeed, unpublished results are beginning to trickle in that the
whiteboard squiggles of a generation of theorists describing their
speculations on a theory called *supersymmetry* may soon be erased
by data from the LHC. Although we need to wait and see, such a
result would provide a dose of humility to those who think they
can infer reality by their thoughts alone, as well as an impetus to
explore more unorthodox approaches.

The application of Platonic reality to physics is fraught with
problems. First, theories are notoriously temporary. We can never
know if quantum field theory will someday be replaced with
another more powerful theory that makes no mention of fields (or
particles, for that matter). Second, as with all physical theories,
quantum field theory is a model—a human invention. We test our
models to find out if they work; but we can never be sure they correspond to "reality." That's metaphysics. If there were an empirical
way to determine ultimate reality, it would be physics, not metaphysics. Third, quantum fields all have quanta that we associate
with the so-called elementary particles.

In relativistic quantum field theory, which is the fundamental
mathematical theory of particle physics and the basis of the standard model, each quantum field has an associated particle called

the quantum of the field. These are the elementary particles of the standard model. The photon is the quantum of the electromagnetic field. The Higgs boson is the quantum of the Higgs field. The electron is the quantum of the Dirac field. I know of no proven example where a quantum field exists without its quantum. Particles are just as much building blocks of our theories as fields. In fact, they are the same building blocks. There are no exceptions. For every field, we have a particle; for every particle, we have a field. So it is incorrect to think that field and particle exist as separate realities. We do not have a field-particle duality. We have, as Pullman says, a field-particle unity.

Please note that the elementary particles of the standard model are not to be thought of as classical objects like billiard balls; they obey all the rules of quantum mechanics. For example, as Feynman showed back in 1948, electrons can zigzag back and forth in space-time and thereby appear many places at the same time. This is usually called *nonlocality*, but a better term is *multilocality*. In this picture, the electron never moves faster than the speed of light.

WAVE-PARTICLE DUALITY

How does this relate to the so-called wave-particle duality that you read about in books on quantum mechanics? The authors often write, "An object is either a particle or a wave, depending on what you decide to measure." This is very misleading and has led to the widespread belief that quantum mechanics shows that human consciousness has the ability to control reality, namely, to decide whether an object is a particle *or* a wave. I have confronted these claims in two previous books and shown them to be specious.[8]

For those who have not moved beyond nonrelativistic Schrödinger wave mechanics, the wave picture provides a perfectly good model to compute quantum effects without having to think about what is doing the waving. To nuclear and particle physicists who must deal with higher-energy phenomena, relativistic quantum mechanics and quantum field theory provide the tools for their calculations without having to think about which is more real—fields or particles. Both are fully materialistic and constitute triumphs for atomism.

In short, quantum physics has not done away with matter. When you kick a rock, it still kicks back. And when you kick an electron, it kicks back.

REDUCTION AND EMERGENCE

Far from demonstrating the existence of a holistic universe in which everything is intimately connected to everything else, relativity and quantum mechanics (and the standard model that was built upon their foundation) confirmed that the universe is reducible to discrete, separated parts. No continuous aether exists throughout the universe. Light is not some vibrating wave in a cosmic medium but is best modeled as a beam of photons streaming through the void. Electricity is not some continuous field moving from place to place but is best modeled as a beam of electrons streaming through the void. (A copper wire is mostly void.)

Nevertheless, strong dissenting voices can be found among scientists in other fields, as well as religious apologists and New Age gurus, that claim phenomena exist that cannot be reduced to elementary particle physics. These phenomena are called *emergent*.[9]

The dissenters correctly point out that the equations of particle physics cannot be used to derive the properties of most complex systems of atoms and molecules, such as the structure of DNA or the behavior of an ant colony. I say "most" because there are several examples from physics where collective properties can be derived from basic particle interactions. This includes all the laws of classical thermodynamics that were originally discovered from macroscopic observation of mundane machines such as heat engines. For example, the ideal gas equation is easily derived in freshman physics from simple particle collisions. The same is true for fluid mechanics, while solids are now also well understood using quantum physics. And, it needs to be pointed out, the physics of classical waves such as sound can also be derived from particle mechanics. There is no wave-particle duality in either classical mechanics or quantum mechanics.

Nevertheless, it is true that the principles that describe most complex systems cannot be derived from particle physics. Chemists, biologists, neuroscientists, sociologists, political scientists, and economists never need to learn about quarks and gauge bosons but develop their own descriptions based on their own observations. You don't need to consult with a quantum mechanic to fix your car. Yet the car and automotive mechanics are still made of material particles.

However, these facts do not imply that emergent properties cannot arise from particle interactions alone—that some new laws of nature operating on the collective scale must come into play. While accepting that we have "bottom-up causality" in which collective properties of complex objects emerge from the interactions of their constituent particles, many authors wishfully seek out an additional "top-down causality," operating in the opposite direction, in which some overarching, universal force acts down to control the behavior of particles at the lowest levels.

Some scientists imagine a natural force akin to Aristotle's principle of final cause.[10] Theologians imagine God as that top-down force.[11] In both cases, we see a strong desire to find purpose in the universe that is absent in both ancient and modern atomism. However, if top-down causality existed, we would expect to see some evidence at all levels of complexity—in social systems, brain processes, biological mechanisms, chemical reactions, and on down to subatomic events. The fact is, we don't. Based on the absence of evidence that should be there but is not, we can rule out beyond a reasonable doubt the existence of top-down emergence. The universe remains reducible to it parts, just as the ancient atomists predicted.

THE ROLE OF CHANCE

Physics from Aristotle to Pierre-Simon Laplace (1749–1827) was marked by the notion of logical principles governing the behavior of matter, with every outcome predetermined by what went on before. Atomism also envisaged atoms moving in a determined direction, "downward" toward some center of attraction, which need not necessarily be the center of Earth. However, on top of that motion was a random "swerve" that causes atoms to deviate from their paths so that they could come in contact with other atoms and stick together to form objects.

Although random motion had been utilized in the nineteenth century by Maxwell, Boltzmann, Gibbs, and others to derive the macroscopic (emergent) laws of thermodynamics from the statistical mechanics of atomic motions, the physics governing these motions was still classical and, thus, deterministic. It was assumed that the laws of Newtonian mechanics ultimately predetermined atomic motions. However, because of the large number of atoms involved, physicists had no hope of calculating individual motions directly, nor really any need for knowing them. By using statistical methods, they were able to calculate various average properties of the whole system, such as pressure and temperature, that sufficed for all practical applications.

With the rise of quantum mechanics and the Heisenberg uncertainty principle in the early twentieth century, the "true" randomness inherent in the motion of all bodies became built into the structure of physics.

Physicists had (almost) no trouble giving up the determinism of the Newtonian world machine. The one exception was Bohmian quantum mechanics, which still has some supporters but is largely dismissed by most physicists because of its implication of superluminal connections for which no evidence exists.[12] However, theists cannot abide events happening outside the supervision of God. We see this in the rejection of Darwinian evolution by virtually all Christians and Muslims. Oh, Catholics and moderate Protestants claim they accept evolution, but only if it is God-guided. But Darwinian evolution is unguided. The Darwinian conclusion that humanity is an accident rather than a divine creation contradicts fundamental teachings of all the major religions on Earth.

Similarly, the role of randomness in the physical universe is irreconcilable with the theistic belief that God created and sustains the universe. Some Christian apologists have tried to make the best of this by suggesting that God operates through the medium of chance.[13] Others have attempted to find a way for God to act in the universe without violating the laws of physics by calling upon quantum mechanics and chaos theory. In my 2009 book, *Quantum Gods*, I show why these arguments all fail.[14]

Put simply, I admit that the god these scholars propose is not ruled out by any current empirical or theoretical knowledge. However, it simply isn't the Judaic-Christian-Islamic God but rather a special deist god who creates the universe, rolls the dice to create complete randomness, and then leaves it alone. It's the god Einstein refused to accept, the "god who places dice." Furthermore, it's not the Enlightenment deist God, who created the fully deterministic Newtonian World Machine. Also, note that by starting up the universe in total chaos, the god who plays dice leaves no memory of its existence or intentions. In short, we humans have no way to infer its existence except to say it is the only god consistent with the data.

Although the god who plays dice is possible, it is certainly not needed. The universe can toss its own dice, which leads us nicely into a discussion of the atomists' atheistic view of an infinite, eternal cosmos and how it contrasts with the traditional religious teaching of a finite, created universe.

THE COSMOS

The multiverse, as we now conceive it, is unlimited in space and eternal in time. It had no beginning and will have no end. It did not come from nothing. It did not have to come from anything because it always was. The discovery of the big bang encouraged theologians to argue that science supports a supernatural creation a finite time ago, although far more distant in the past and far different in details than any scriptural account of creation ever imagined. What's more, the universe revealed by astronomy is incredibly larger than what was supposedly revealed in sacred books.

Today, physicists and cosmologists have shown that while *our* universe undoubtedly began with the big bang 13.7 billion years ago, mathematically precise models based on general relativity and quantum mechanics suggest that it may have quantum-tunneled from an earlier universe. Furthermore, these models also imply that our universe is just one of an enormous, if not unlimited, number of other universes in a vast multiverse.

Now, theologians have argued that this proposal is unscientific. First, they say we have no evidence for other universes, which is true, but such evidence is not beyond our reach. The presence of another universe is in principle detectable by its gravitational effect on ours, for example, by producing an anisotropy in the cosmic background radiation.

Second, the multiple-universe hypothesis is said to be non-parsimonious, in violation of Ockham's razor. This is untrue. Ockham's razor applies to the hypotheses of a theory, not the number of entities in the theory. The atomic theory multiplied the number of entities physicists had to deal with by a trillion-trillion, yet it was more parsimonious than previous theories because it involved fewer hypotheses. It takes an additional hypothesis to assume that only one universe exists when the current theory, with fewer hypotheses, predicts many. That is, it is the single-universe theory that violates Ockham's razor.

The multiverse is also unlimited in spatial extent, again in agreement with the atomist picture. Furthermore, our visible universe, all 100 billion galaxies, may be only a grain of sand on the Sahara of galaxies that arose from the original big bang and now continues far beyond our light horizon. While the ancient atomists, of course, had none of this knowledge, their original intuition continues to be right on the mark as we progress in our knowledge of the cosmos.

THE MIND

The notion that only atoms are real contradicted the teaching of Aristotle, who regarded sense impressions as constituting an intrinsic reality.[15] To Aristotle, redness is a real property of a tomato and sweetness is a real property of sugar, while the atomists maintained these qualities are "conventions"—mental perceptions that describe human reactions to the messages of their senses.

Modern cognitive science fully supports the atomists' view. The intrinsic properties of elementary particles, such as mass and energy, are now referred to as primary properties. Objectively observed secondary properties, such as color and wetness, arise from the interactions of particles and are not intrinsic to the objects that are assigned these properties. The subjective sense impressions that arise from secondary properties I have termed secondary *qualities*.

I have not said much about the human mind in this book, which has focused on the physical evidence that supports the model of atoms and the void. I have asserted (many times now) that all we observe with our senses and scientific instruments can at least plausibly, and in most cases explicitly, be explained without the inclusion of immaterial elements. In this, I expect to get disagreement from dualists who will say that because we still don't have a consensus material explanation of consciousness, the door is still open for there to be a nonmaterial component to the human mind.

I claim I can make a similar statement about those phenomena conventionally labeled "mental": all we observe concerning mental phenomena with our senses and scientific instruments can at least plausibly, and in most cases explicitly, be explained without the inclusion of immaterial elements. As I covered in detail in *God and the Folly of Faith*, neuroscience has provided convincing evidence that the mind is what the brain does, and the brain is fully material.[16] Furthermore, our current knowledge of the brain is consistent with atomism. Indeed, neurons act, in a way, as the "atoms" of the brain.

Of course, debate still rages on the nature of conscious experience, the *qualia* mentioned in chapter 3. But the very subjective nature of that experience makes it difficult to even talk about it scientifically. About the only objective data we have is the correlation between subject reports on types of such experiences and the observed increase in activities in local areas of the brain, along with other physiological measurements.

Models that introduce immaterial elements into the human cognitive process have failed to produce any supportive evidence comparable to materialistic brain science. That would seem to make a pretty good case for the physical source of conscious experiences even though, as yet, there is no received philosophical/scientific explanation of how the brain gives rise to consciousness. So far, brain science, which is very much based on atomism, has been making steady progress in explaining the features of consciousness. Dualism has made none.

NO HIGHER POWER

No doubt, many religious apologists will argue that nothing in the success of the atomic theory of matter, as exemplified by modern elementary particle physics and cosmology, is necessarily inconsistent with the belief in a "higher power." However, we have seen throughout this book that the atomism of Democritus, Epicurus, and Lucretius was very closely based on a worldview in which no such higher power exists. The total absence of empirical facts and theoretical arguments to support the existence of any component to reality other than atoms and the void can be taken as proof beyond a reasonable doubt that such a component is nowhere to be found. The scientific triumph of atomism represents a philosophical triumph for the recognition by the ancient atomists that the world can be understood without calling upon any forces from outside the world—no wood sprites, no fairies, no angels, no devils, no gods or spirits of any sort.

NOTES

1. ANCIENT ATOMISM

1. Eugene O'Connor, *The Essential Epicurus: Letters, Principal Doctrines, Vatican Sayings, and Fragments* (Amherst, NY: Prometheus Books, 1993), p. 21.

2. Andrew Pyle, *Atomism and Its Critics: Problem Areas Associated with the Development of the Atomic Theory of Matter from Democritus to Newton* (Bristol, UK: Thoemmes Press, 1995), p. xi.

3. Ibid.

4. Alex Rosenberg, *Darwinian Reductionism; or, How to Stop Worrying and Love Molecular Biology* (Chicago; London: University of Chicago Press, 2006), p. 12.

5. Victor J. Stenger, *God and the Folly of Faith: The Incompatibility of Science and Religion* (Amherst, NY: Prometheus Books, 2012), pp. 205–18.

6. Bernard Pullman, *The Atom in the History of Human Thought: A Panoramic Intellectual History of a Quest That Has Engaged Scientists and Philosophers for 2,500 Years* (Oxford; New York: Oxford University Press, 1998), pp. 32–37; Geoffrey S. Kirk, John E. Raven, and Malcolm Schofield, *The Presocratic Philosophers: A Critical History with a Selection of Texts*, 2nd ed. (Cambridge; New York: Cambridge University Press, 1983), pp. 402–33.

7. Ibid., pp. 31–32.

8. David N. Sedley, *Creationism and Its Critics in Antiquity* (Berkeley: University of California Press, 2007), pp. 233–34.

9. Kirk, Raven, and Schofield, *Presocratic Philosophers*, p. 410.

10. Leon M. Lederman and Dick Teresi, *The God Particle: If the Universe Is the Answer, What Is the Question?* (1993; repr., Boston: Houghton Mifflin, 2006), pp. 32–58.

11. Pullman, *Atom in the History of Human Thought*, pp. 78–79.

12. Ibid., pp. 79–80.

13. Ibid., pp. 81–84.

14. O'Connor, *Essential Epicurus*.

15. Tim O'Keefe, "Epicurus (341–271 BCE)," *Internet Encyclopedia of Philosophy*,

last modified July 11, 2005, http://www.iep.utm.edu/epicur/#SSH3c.ii (accessed October 26, 2011).

16. Richard Carrier, private correspondence; see also Richard Carrier, "Christianity Was Not Responsible for Modern Science," in *The Christian Delusion: Why Faith Fails*, ed. John W. Loftus (Amherst, NY: Prometheus Books, 2010), pp. 396–419.

17. A. E. Stallings and Richard Jenkyns, *Lucretius: The Nature of Things* (London; New York: Penguin, 2007).

18. Mary Gallagher, "Dryden's Translation of Lucretius," *Huntington Library Quarterly* 28, no. 1 (1964): 19–29.

19. Ibid.

20. Stallings and Jenkyns, *Lucretius*.

21. Jim Holt, "Physicists, Stop the Churlishness," *New York Times*, June 8, 2012.

22. Richard P. Feynman, Robert B. Leighton, and Matthew L. Sands, *The Feynman Lectures on Physics*, new millennium ed. (New York: Basic Books, 2011).

23. Gaston Bachelard, *Les Intuitions Atomistiques [The Atomistic Intuitions]: Essai de Classification*, 2nd ed. (Paris: J. Vrin, 1975).

24. Richard Carrier, "Predicting Modern Science: Epicurus vs. Mohammed," Secular Web, June 22, 2004, http://www.infidels.org/kiosk/article362.html (accessed July 22, 2012).

25. Pullman, *Atom in the History of Human Thought*, pp. 64–66.

26. Ibid., p. 68.

27. Stephen Greenblatt, *The Swerve: How the World Became Modern* (New York: W. W. Norton, 2011), p. 189.

28. Pullman, *Atom in the History of Human Thought*, p. 70.

2. ATOMS LOST AND FOUND

1. Saint Augustine (Bishop of Hippo) and Marcus Dods, *The City of God* (New York: Modern Library, 1993).

2. Bernard Pullman, *The Atom in the History of Human Thought: A Panoramic Intellectual History of a Quest That Has Engaged Scientists and Philosophers for 2,500 Years* (Oxford; New York: Oxford University Press, 1998), pp. 89–94.

3. Gerald Holton, *Victory and Vexation in Science: Einstein, Bohr, Heisenberg, and Others* (Cambridge, MA: Harvard University Press, 2005), p. 27. Selections translated from Werner Heisenberg, *Der Teil und das Ganze* (Munich: R. Piper, 1969), p. 49.

4. Pullman, *Atom in the History of Human Thought*, p. 93.

5. This section is based on ibid., chaps. 9, 10, and 11.

6. Moses Maimonides et al., *The Guide of the Perplexed* (Indianapolis, IN: Hackett, 1995).

7. For an excellent survey, see Jim al-Khalili, *The House of Wisdom: How Arabic Science Saved Ancient Knowledge and Gave Us the Renaissance* (New York: Penguin Press, 2011).

8. I am following al-Khalili here in using the word *Arabic* rather than *Arab* or *Islamic* to describe the science and philosophy of that region because not everyone involved was either an Arab or a Muslim; however, Arabic was the common language.

9. Pullman, *Atom in the History of Human Thought*, p. 114.

10. Alison Brown, *The Return of Lucretius to Renaissance Florence* (Cambridge, MA: Harvard University Press, 2010).

11. Stephen Greenblatt, *The Swerve: How the World Became Modern* (New York: W. W. Norton, 2011).

12. Which is why officially the twenty-third pope named John was Angelo Giuseppe Roncalli, who died in 1963. Actually, Roncalli was the twenty-second pope since there was no John XX.

13. Greenblatt, *Swerve*, p. 49.

14. Ibid., p. 32.

15. Ibid., p. 52.

16. The full manuscript can be viewed online at http://teca.bmlonline.it/TecaViewer/index.jsp?RisIdr=TECA0000416254&keyworks=Plut.35.30 (accessed March 31, 2012).

17. Greenblatt, *Swerve*, p. 184.

18. Ibid., p. 199.

19. *A Man for All Seasons*, a play by Robert Bolt, was first performed in London on July 1, 1960.

20. Greenblatt, *Swerve*, pp. 227–31.

21. Ibid., p. 237.

22. Barry Brundell, *Pierre Gassendi: From Aristotelianism to a New Natural Philosophy* (Dordrecht, Neth.; Boston; Norwell, MA: D. Reidel, 1987); Lynn Sumida Joy, *Gassendi, the Atomist: Advocate of History in an Age of Science* (Cambridge; New York: Cambridge University Press, 1987); Antonia LoLordo, *Pierre Gassendi and the Birth of Early Modern Philosophy* (New York: Cambridge University Press, 2007).

23. Of course, Earth is not in uniform motion but rotates on its axis and turns as it orbits the sun. But these accelerations are not noticeable to most residents.

24. Saul Fisher, "Pierre Gassendi," *Stanford Encyclopedia of Philosophy*, last modified December 15, 2009, http://plato.stanford.edu/entries/gassendi/ (accessed November 6, 2011).

25. Ibid.

26. Wes Wallace, "The Vibrating Nerve Impulse in Newton, Willis and Gassendi: First Steps in a Mechanical Theory of Communication," *Brain and Cognition* 51 (2003): 66–94.

3. ATOMISM AND THE SCIENTIFIC REVOLUTION

1. See note 8 in chapter 2.

2. Victor J. Stenger, *God and the Folly of Faith: The Incompatibility of Science and Religion* (Amherst, NY: Prometheus Books, 2012), pp. 74–76.

3. Pierre Maurice Marie Duhem, *To Save the Phenomena: An Essay on the Idea of Physical Theory from Plato to Galileo* (Chicago: University of Chicago Press, 1969).

4. James Hannam, *God's Philosophers: How the Medieval World Laid the Foundations of Modern Science* (London: Icon, 2009).

5. Jonathan Kirsch, *God against the Gods: The History of the War between Monotheism and Polytheism* (New York: Viking Compass, 2004).

6. Stenger, *God and the Folly of Faith*, p. 71.

7. David C. Lindberg, *The Beginnings of Western Science: The European Scientific Tradition in Philosophical, Religious, and Institutional Context, Prehistory to A.D. 1450*, 2nd ed. (Chicago: University of Chicago Press, 2007), pp. 308–309.

8. Edith Dudley Sylla, *The Oxford Calculators and the Mathematics of Motion, 1320–1350: Physics and Measurement by Latitudes* (New York: Garland, 1991).

9. Lindberg, *Beginnings of Western Science*, pp. 300–13.

10. Ibid., pp. 359–64.

11. Alexandre Koyré, *Metaphysics and Measurement* (Yverdon, Switz.; Langhorne, PA: Gordon and Breach Science, 1992), pp. 20–21.

12. Ibid.

13. See Stenger, *God and the Folly of Faith*, p. 83 and references therein.

14. Michael White, *Isaac Newton: The Last Sorcerer* (Reading, MA: Addison-Wesley, 1997), pp. 190–91.

15. Ibid., p. 238 and references therein.

16. At speeds near the speed of light, the equation for momentum is more complicated, but this need not concern us here.

17. As quoted in Bernard Pullman, *The Atom in the History of Human Thought: A Panoramic Intellectual History of a Quest That Has Engaged Scientists and Philosophers for 2,500 Years* (Oxford; New York: Oxford University Press, 1998), p. 139.

18. Lindberg, *Beginnings of Western Science*, p. 365.

19. Pullman, *Atom in the History of Human Thought*, p. 127.

20. Isaac Newton, *Mathematical Principles of Natural Philosophy* (New York: Greenwood Press, 1969).

21. Lawrence Nolan, ed., *Primary and Secondary Qualities: The Historical and Ongoing Debate* (New York: Oxford University Press, 2011), p. 1.

22. Ibid.

23. Alex Byrne and David R. Hilbert, "Are Colors Secondary?" in ibid., pp. 339–61; Barry Maund, "Color Eliminativism," in ibid., pp. 362–85.

24. Pullman, *Atom in the History of Human Thought*, pp. 141–42.

25. Ibid., pp. 142–43.

26. Ibid., pp. 144–45.

27. Ibid., pp. 146–51.

28. Philipp Blom, *A Wicked Company: The Forgotten Radicalism of the European Enlightenment* (New York: Basic Books, 2010); see also Stenger, *God and the Folly of Faith*, pp. 93–94.

29. Julien O. de la Mettrie, *L'Homme Machine* (Lyde, 1748).

30. Pullman, *Atom in the History of Human Thought*, pp. 155–63.

31. Ibid., pp. 166–73.

4. THE CHEMICAL ATOM

1. Ida Freund, *The Study of Chemical Composition. An Account of Its Method and Historical Development, with Illustrative Quotations* (Cambridge: Cambridge University Press, 1904), p. 288.

2. Jim al-Khalili, *The House of Wisdom: How Arabic Science Saved Ancient Knowledge and Gave Us the Renaissance* (New York: Penguin Press, 2011), pp. 49–66.

3. Michael White, *Isaac Newton: The Last Sorcerer* (Reading, MA: Addison-Wesley, 1997), pp. 115–116.

4. Ibid., p. 131.

5. Ibid., p. 109.

6. Ibid., pp. 152–53.

7. Ibid., pp. 153–57.

8. Ian G. Barbour, *Religion and Science: Historical and Contemporary Issues* (San Francisco, CA: HarperSanFrancisco, 1997), p. 24.

9. Norman E. Holden, "History of the Origin of the Chemical Elements and Their Discoverers," National Nuclear Data Center, last modified March 12, 2004, http://www.nndc.bnl.gov/content/elements.html (accessed November 24, 2011).

10. John Dalton, *A New System of Chemical Philosophy* (Manchester; London, 1808).

11. Brewton-Parker College, "History of the Development of the Periodic Table of Elements," April 15, 2010, http://www.bpc.edu/mathscience/chemistry/history_of_the_periodic_table.html (accessed November 26, 2011; site discontinued).

12. Jean-Baptiste Dumas, *Lectures Delivered before the Chemical Society. Faraday Lectures 1869–1928* (London: Chemical Society, 1928).

13. Bernard Pullman, *The Atom in the History of Human Thought: A Panoramic Intellectual History of a Quest That Has Engaged Scientists and Philosophers for 2,500 Years* (Oxford; New York: Oxford University Press, 1998), p. 234.

14. See the appendix "Physics of a Believer" in Pierre Maurice Marie Duhem, *The Aim and Structure of Physical Theory* (Princeton, NJ: Princeton University Press, 1954).

15. Georg Wilhelm Hegel, *Encyclopedia of the Philosophical Sciences in Outline, and Critical Writings*, trans. Steven A. Taubeneck (New York: Continuum, 1990).

16. Pullman, *Atom in the History of Human Thought*, p. 211.

17. Arthur Schopenhauer, *The World as Will and Representation* (New York: Dover, 1958).

18. As quoted in Pullman, *Atom in the History of Human Thought*, p. 217.

5. ATOMS REVEALED

1. From Gibbs's letter accepting the Rumford Medal (1881). Quoted in Alan L. Mackay, *A Dictionary of Scientific Quotations* (Bristol, UK; Philadelphia: A. Hilger, 1991).

2. Vladimir Shiltsev, "Mikhail Lomonosov and the Dawn of Russian Science," *Physics Today* 65, no. 2 (2012): 40–45.

3. David Lindley, *Boltzmann's Atom: The Great Debate That Launched a Revolution in Physics* (New York: Free Press, 2001), p. 3.

4. Thomas Kuhn, "Energy Conservation as an Example of Simultaneous Discovery," in *Critical Problems in the History of Science*, ed. Marshall Clagett (Madison: University of Wisconsin Press, 1969), pp. 321–56.

5. As quoted in Albert L. Lehninger, *Bioenergetics: The Molecular Basis of Biological Energy Transformations*, 2nd ed. (Menlo Park, CA: W. A. Benjamin, 1971), p. 20.

6. Rudolf Clausius, "Ueber die bewegende Kraft der Wärme und die Gesetze, welche sich daraus für die Wärmelehre selbst ableiten lassen," *Annalen der Physik und Chemie* 155, no. 3 (1850): 368–94 and 384.

7. John Tyndall and William Francis, *Scientific Memoirs, Selected from the Transactions of Foreign Academies of Science, and from Foreign Journals. Natural Philosophy* (London: Taylor and Francis, 1853).

8. James Prescott Joule, "On the Rarefaction and Condensation of Air," *Philosophical Magazine, Scientific Papers* 172 (1845).

9. James Prescott Joule, "On the Mechanical Equivalent of Heat," *Philosophical Transactions of the Royal Society of London* 140, no. 1 (1850): 61–82.

10. Julian T. Rubin, "James Prescott Joule: The Discovery of the Mechanical Equivalent of Heat," Following the Path of Discovery, last modified July 2012, http://www.juliantrubin.com/bigten/mechanical_equivalent_of_heat.html (accessed February 4, 2012).

11. The nutritionist Calorie is actually 1,000 physicist calories.

12. William Thomson, "On an Absolute Temperature Scale Founded on Carnot's Theory of the Motive Power of Heat, and Calculated from Regnault's Observations," *Philosophical Magazine* 1 (1848). Reprinted in William Thomson, *Mathematical and Physical Papers* (Cambridge: Cambridge University Press, 1882), pp. 100–106.

13. Lindley, *Boltzmann's Atom*.

14. Bernard Pullman, *The Atom in the History of Human Thought: A Panoramic Intellectual History of a Quest That Has Engaged Scientists and Philosophers for 2,500 Years* (Oxford; New York: Oxford University Press, 1998), p. 202. ·

15. Pascal was very religious and argued that you have everything to win by betting on God, even if you don't believe in him, and everything to lose if you don't. He didn't explain why God would want to spend eternity in a heaven full of liars.

16. Claude Elwood Shannon and Warren Weaver, *The Mathematical Theory of Communication* (Urbana: University of Illinois Press, 1949).

17. Lehninger, *Bioenergetics*, p. 25.

18. W. S. Gilbert and Arthur Sullivan's opera, *H.M.S. Pinafore*, opened in 1878.

19. Ernst Mach, *The Science of Mechanics; a Critical and Historical Exposition of Its Principles*, trans. Thomas J. McCormack (Chicago: Open Court, 1893).

20. Pierre Maurice Marie Duhem, *Thermodynamics and Chemistry: A Non-Mathematical Treatise for Chemists and Students of Chemistry* (New York; London: J. Wiley and Sons; Chapman and Hall, 1903).

21. Pullman, *Atom in the History of Human Thought*, p. 237.

22. Lindley, *Boltzmann's Atom*, pp. 149–60.

23. W. Ostwald, "La Déroute de L'Atomisme Contermporain" [The Disarray of Contemporary Atomism], *Revue Génerale des Sciences Pures et Appliquée*, no. 21 (1895): 953–58.

24. Lindley, *Boltzmann's Atom*, p. 128.

25. Pullman, *Atom in the History of Human Thought*, p. 221.

26. Ibid.

27. Lindley, *Boltzmann's Atom*, p. 169.

28. Ibid., p. 171.

29. Ibid., p. 199.

30. Ibid., p. 201.

31. Ibid., pp. 203–205.

32. Drew Dolgert, "Einstein's Explanation of Brownian Motion," Fowler's Physics Applets, September 22, 1998, http://Galileo.phys.Virginia.EDU/classes/109N/more_stuff/Applets/brownian/applet.html (accessed November 28, 2011).

33. Albert Einstein, R. Fürth, and Alfred Denys Cowper, *Investigations on the Theory of the Brownian Movement* (London: Methuen, 1926).

34. Lindley, *Boltzmann's Atom*, pp. 218–19.

6. LIGHT AND THE AETHER

1. William Davidson Niven, ed., *The Scientific Papers of James Clerk Maxwell*, vol. 2 (Cambridge: Cambridge University Press, 1890).

2. David C. Lindberg, *The Beginnings of Western Science: The European Scientific Tradition in Philosophical, Religious, and Institutional Context, Prehistory to A.D. 1450*, 2nd ed. (Chicago: University of Chicago Press, 2007), pp. 313–14.

3. Michael White, *Isaac Newton: The Last Sorcerer* (Reading, MA: Addison-Wesley, 1997), pp. 170–71.

4. H. Turnbull, ed., *The Correspondence of Isaac Newton*, vol. 1 (Cambridge: Cambridge University Press, 1959), pp. 110–11.

5. White, *Isaac Newton*, p. 182.

6. Ibid.

7. Turnbull, *Correspondence of Isaac Newton*, p. 416.

8. White, *Isaac Newton*, p. 187.

9. Turnbull, *Correspondence of Isaac Newton*, pp. 171–73.

10. White, *Isaac Newton*, p. 208.

11. Michael Faraday, "On the Physical Lines of Magnetic Force," in *Great Books of the Western World. Encyclopaedia Britannica, Inc., in Collaboration with the University of Chicago*, vol. 45, ed. Robert Maynard Hutchins (Chicago: W. Benton, 1952), p. 530.

12. Ibid., p. 759.

13. I have added a subscript to Einstein's famous equation because I want to use E for the total energy.

14. Leon M. Lederman and Christopher T. Hill, *Symmetry and the Beautiful Universe* (Amherst, NY: Prometheus Books, 2004), chaps. 3 and 5.

15. Nina Byers, "E. Noether's Discovery of the Deep Connection between Symmetries and Conservation Laws" (paper presented at Israel Mathematical Conference, Bar Ilan University, Tel Aviv, Israel, December 2–3, 1996), http://www .physics.ucla.edu/~cwp/articles/noether.asg/noether.html (accessed February 20, 2009).

7. INSIDE THE ATOM

1. Abraham Pais, *The Genius of Science: A Portrait Gallery* (Oxford; New York: Oxford University Press, 2000), p. 24.

2. Victor Stenger, *The Fallacy of Fine-Tuning: Why the Universe Is Not Designed for Us* (Amherst, NY: Prometheus Books, 2011).

3. David Lindley, *Boltzmann's Atom: The Great Debate That Launched a Revolution in Physics* (New York: Free Press, 2001), p. 115.

4. Ibid., p. 175.

5. For a circular orbit, the angular momentum is mvr, where m is the mass of the electron, v is its speed, and r is the radius of the orbit.

6. Olaf Nairz, Markus Arndt, and Anton Zeilinger, "Quantum Interference Experiments with Large Molecules," *American Journal of Physics* 71, no. 10 (2003): 319–25.

7. This is because their masses are so large. While the speeds of macroscopic objects can be very low, they are never exactly zero and mv is always very much greater than h.

8. P. A. M. Dirac, *The Principles of Quantum Mechanics*, 4th ed. (1930; repr., Oxford: Clarendon Press, 1958).

9. Ibid., p. 80.

10. D. J. Bohm and B. J. Hiley, "The De Broglie Pilot Wave Theory and the Further Development of New Insights Arising out of It," *Foundations of Physics* 12, no. 10 (1982): 1001–16.

11. David Bohm and B. J. Hiley, *The Undivided Universe: An Ontological Interpretation of Quantum Theory* (London; New York: Routledge, 1993).

12. Victor J. Stenger, *The Comprehensible Cosmos: Where Do the Laws of Physics Come From?* (Amherst, NY: Prometheus Books, 2006), pp. 240–41.

8. INSIDE THE NUCLEUS

1. Interview about the Trinity explosion, the first nuclear bomb test. Broadcast as part of the television documentary *The Decision to Drop the Bomb*, produced by Fred Freed (1965; NBC White Paper).

2. In principle, potential energies should be included, but these are generally not important. They result from any outside forces, but the reactions generally occur in isolation from these forces, other than gravity, which affects both sides of the reaction equally.

3. R. Svoboda and K. Gordan, "Neutrinos in the Sun," Astronomy Picture of the Day, June 5, 1998, http://apod.nasa.gov/apod/ap980605.html (accessed January 11, 2012).

4. World Health Organization, "Air Quality and Health: Fact Sheet no. 313," last modified September 2011, http://www.who.int/mediacentre/factsheets/fs313/en/index.html (accessed January 2, 2012).

5. Victor J. Stenger, *God and the Folly of Faith: The Incompatibility of Science and Religion* (Amherst, NY: Prometheus Books, 2012), pp. 313–20.

6. Nuclear Energy Institute, "Nuclear Energy around the World," 2012, http://www.nei.org/resourcesandstats/nuclear_statistics/worldstatistics/ (accessed January 17, 2013).

7. Richard G. Hewlett and Jack M. Holl, *Atoms for Peace and War, 1953–1961: Eisenhower and the Atomic Energy Commission* (Berkeley: University of California Press, 1989), pp. 192–95.

8. International Atomic Energy Agency, World Health Organization, and United Nations Development Agency, "Chernobyl: The True Scale of the Accident," press release, September 5, 2005, http://www.iaea.org/newscenter/focus/chernobyl/pdfs/pr.pdf (accessed December 29, 2011).

9. Daniel Yergin, *The Quest: Energy, Security and the Remaking of the Modern World* (New York: Penguin Press, 2011), p. 377.

10. Ibid., pp. 372–73.

11. Ibid., p. 403.

12. Robert Hargraves and Ralph Moir, "Liquid Fluoride Thorium Reactors: An Old Idea in Nuclear Power Gets Reexamined," *American Scientist* 98, no. 4 (2010): 304–13.

13. Robert Hargraves and Ralph Moir, "Liquid Fuel Nuclear Reactors," *Physics & Society* 40, no. 1 (2011): 6–10.

14. Yergin, *Quest*, p. 403.

15. Ibid., p. 407.

16. Robert Hargraves, "Aim High: Using Thorium to Address Environmental

Problems," YouTube video, 59:50, from Google Tech Talk, posted by "GoogleTech-Talks," May 26, 2009, http://www.youtube.com/watch?v=VgKfS74hVvQ (accessed January 2, 2012).

17. Richard Martin, *Superfuel: Thorium, the Green Energy Source for the Future* (New York: Palgrave Macmillan, 2012).

9. QUANTUM FIELDS

1. Richard P. Feynman, *QED: The Strange Theory of Light and Matter* (Princeton, NJ: Princeton University Press, 1985), p. 15.

2. Hans A. Bethe, "The Electromagnetic Shift of Energy Levels," *Physical Review* 72, no. 4 (1947): 339–41.

3. Julian Schwinger, "On Quantum-Electrodynamics and the Magnetic Moment of the Electron," *Physical Review* 73, no. 4 (1948): 416–17.

4. Richard Feynman, "Space-Time Approach to Quantum Electrodynamics," *Physical Review* 76, no. 6 (1949): 769–89.

5. Sin-Itiro Tomonaga, "On a Relativistically Invariant Formulation of the Quantum Theory of Wave Fields," *Progress in Theoretical Physics* 1, no. 2 (1946): 27–42; free copy available at http://ptp.ipap.jp/link?PTP/1/27/ (accessed January 4, 2012.)

6. For a complete history of QED, see Silvan S. Schweber, *QED and the Men Who Made It: Dyson, Feynman, Schwinger, and Tomonaga* (Princeton, NJ: Princeton University Press, 1994).

7. Freeman J. Dyson, "The Radiation Theories of Tomonaga, Schwinger, and Feynman," *Physical Review* 75, no. 3 (1949): 486–502.

8. David Kaiser, "Physics and Feynman's Diagrams," *American Scientist* 93 (2005): 156–65.

9. For a further discussion of the point, see Victor J. Stenger, *Timeless Reality: Symmetry, Simplicity, and Multiple Universes* (Amherst, NY: Prometheus Books, 2000).

10. Ernst Stückelberg, "La mécanique du point matériel en théorie de relativité et en théorie des quanta," *Helvetica Physica Acta* 15 (1942): 23–37.

11. Richard Feynman, "The Theory of Positrons," *Physical Review* 76 (1949): 749–59.

12. Stückelberg, "La mécanique."

13. F. Mandl and G. Shaw, *Quantum Field Theory*, rev. ed. (Chichester, UK; New York: Wiley, 1993), pp. 7–10. This book is the classical textbook for quantum field theory.

10. THE RISE OF PARTICLE PHYSICS

1. March 2007 TED talk on beauty and truth in physics.

2. I show this mathematically in Victor J. Stenger, *The Comprehensible Cosmos: Where Do the Laws of Physics Come From?* (Amherst, NY: Prometheus Books, 2006), pp. 266–68.

3. Hideki Yukawa, "On the Interaction of Elementary Particles," *Progress in Theoretical Physics* 17 (1935): 48–56; English translation available at http://web.ihep.su/dbserv/compas/src/yukawa35/eng.pdf (accessed February 20, 2012).

4. Geoffrey F. Chew, "S-Matrix Theory of Strong Interactions without Elementary Particles," *Reviews of Modern Physics* 34, no. 3 (1962): 394–401.

5. Fritjof Capra, *The Tao of Physics: An Exploration of the Parallels between Modern Physics and Eastern Mysticism* (Berkeley, CA: Shambhala, 1975).

6. Ibid., pp. 276–77.

7. Marilyn Ferguson, *The Aquarian Conspiracy: Personal and Social Transformation in the 1980s* (Los Angeles; New York: J. P. Tarcher, 1980); Robert Basil, ed., *Not Necessarily the New Age: Critical Essays* (Amherst, NY: Prometheus Books, 1988).

8. Victor J. Stenger, *The Unconscious Quantum: Metaphysics in Modern Physics and Cosmology* (Amherst, NY: Prometheus Books, 1995).

9. Victor J. Stenger, *Quantum Gods: Creation, Chaos, and the Search for Cosmic Consciousness* (Amherst, NY: Prometheus Books, 2009).

11. THE DREAMS THAT STUFF IS MADE OF

1. Steven Weinberg, *The First Three Minutes: A Modern View of the Origin of the Universe* (New York: Basic Books, 1977).

2. V. E. Barnes et al., "Observation of a Hyperon with Strangeness Minus Three," *Physical Review Letters* 12, no. 8 (1964): 204–206.

3. J. Berginger et al., "The Review of Particle Physics," *Physical Review* D86 (2012): 010001.

4. Super-Kamiokande Collaboration, "Evidence for Oscillation of Atmospheric Neutrinos," *Physical Review Letters* 81 (1998): 1562–67.

5. Victor J. Stenger, "Neutrino Oscillations in DUMAND" (paper presented at the Neutrino Mass Mini-Workshop, Cable, WI, 1980); a scanned copy of the original paper can be found at http://www.colorado.edu/philosophy/vstenger/Telemark.pdf (accessed July 1, 2012).

6. Although Noether's theorem specifically applied to space-time symmetries, it can be extended to other dimensions.

7. I discuss gauge transformations in detail in Victor J. Stenger, *The Comprehensible Cosmos: Where Do the Laws of Physics Come From?* (Amherst, NY: Prometheus Books, 2006).

8. A complex number can be written $c = a + ib$, where $i = \sqrt{-1}$ and where a and b are real numbers. It can also be written $c = A\exp(\phi)$, where A is the amplitude, ϕ is the phase, and exp is the exponential function.

9. Ian Sample, *Massive: The Missing Particle That Sparked the Greatest Hunt in Science* (New York: Basic Books, 2012).

10. See Frank Wilczek, "QCD Made Simple," *Physics Today* 53, no. 8 (2000): 22–28.

11. F. Englert and R. Brout, "Broken Symmetry and the Mass of Gauge Vector Bosons," *Physical Review Letters* 13 (1964): 321; Peter W. Higgs, "Broken Symmetries and the Masses of Gauge Bosons," *Physical Review Letters* 13 (1964): 508; G. G. Guralnik, C. R. Hagen, and T. W. Kibble, "Global Conservation Laws and Massless Particles," *Physical Review Letters* 13 (1964): 585.

12. Leon M. Lederman and Dick Teresi, *The God Particle: If the Universe Is the Answer, What Is the Question?* (1993; repr., Boston: Houghton Mifflin, 2006).

13. Sample, *Massive*, p. 123.

14. The reader interested in looking at a short summary of the mathematics can go to Stenger, *Comprehensible Cosmos*, pp. 268–72.

15. Sample, *Massive*.

16. Steven Weinberg, *Dreams of a Final Theory* (New York: Pantheon Books, 1992).

17. See P. W. Anderson, "More Is Different," *Science* 177, no. 4047 (1972): 393–96.

18. Sample, *Massive*, p. 114.

19. Ibid., p. 121.

20. ATLAS Collaboration, "Observation of a New Particle in the Search for the Standard Model Higgs Boson with the ATLAS Detector at the LHC," *Physics Letters B* 716, no. 1 (2012): 1–29.

21. CMS Collaboration, "Observation of a New Boson at a Mass of 125 Gev with the CMS Experiment at the LHC," *Physics Letters B* 716, no. 1 (2012): 30–61.

22. C. T. H. Davies et al., "High-Precision Lattice QCD Confronts Experiment," *Physical Review Letters* 92, no. 2 (2004): 022001–022005.

23. Lillian Hoddeson, *The Rise of the Standard Model: Particle Physics in the 1960s and 1970s* (New York: Cambridge University Press, 1997).

24. Howard Georgi and Sheldon Glashow, "Unity of All Elementary Particle Forces," *Physical Review Letters* 32 (1974): 438–41.

25. H. Nishino et al., "Search for Proton Decay via $p \rightarrow e^+ + \pi^0$ and $p \rightarrow \mu^+ + \pi^0$ in a Large Water Cherenkov Detector," *Physical Review Letters* 102 (2009): 141801.

12. ATOMS AND THE COSMOS

1. Georges Lemaître, "Un univers homogène de masse constante et de rayon croissant rendant compte de la vitesse radiale des nébuleuses extra-galactiques" [A Homogeneous Universe of Constant Mass and Growing Radius Accounting for the Radial Velocity of Extragalactic Nebulae], *Annales de la Société Scientifique de Bruxelles* 47 (1927): 49.

2. Pope Pius XII, "The Proofs for the Existence of God in the Light of Modern Natural Science" (address to the Pontifical Academy of Sciences, November 22, 1951). Reprinted as "Modern Science and the Existence of God," *Catholic Mind* 49 (1972): 182–92.

3. Marcia Bartusiak, *The Day We Found the Universe* (New York: Pantheon Books, 2009).

4. Our nearest neighbor galaxy (excluding the Magellanic Clouds, which are small satellites of the Milky Way) is Andromeda, which is currently 2.6 million light-years away. Unlike most other galaxies, it is moving toward us, and recent observations with the Hubble Space Telescope confirm that it will collide with the Milky Way in 4 billion years. However, since both galaxies are mostly empty space, they should just pass through each other largely unscathed.

5. Donald Goldsmith, *Einstein's Greatest Blunder? The Cosmological Constant and Other Fudge Factors in the Physics of the Universe* (Cambridge, MA: Harvard University Press, 1995).

6. Demos Kazanas, "Dynamics of the Universe and Spontaneous Symmetry Breaking," *Astrophysical Journal* 241 (1980): L59–L63; Alan H. Guth, "Inflationary Universe: A Possible Solution to the Horizon and Flatness Problems," *Physical Review D* 23, no. 2 (1981): 347–56; Andrei D. Linde, "A New Inflationary Universe Scenario: A Possible Solution of the Horizon, Flatness, Homogeneity, Isotropy and Primordial Monopole Problems," *Physics Letters B* 108 (1982): 389; Alan H. Guth, *The Inflationary Universe: The Quest for a New Theory of Cosmic Origins* (Reading, MS: Addison-Wesley, 1997).

7. George Smoot and Keay Davidson, *Wrinkles in Time: Witness to the Birth of the Universe* (New York: Harper Perennial, 2007).

8. N. Jarosik et al., "Seven-Year Wilkinson Microwave Anisotropy Probe (WMAP). Observations: Sky Maps, Systematic Errors, and Basic Results," *Astrophysical Journal* 192 (2011): 14.

9. See NASA's website, "Wilkinson Microwave Anisotropy Probe," last modified December 21, 2012, http://map.gsfc.nasa.gov/ (accessed January 16, 2013).

10. Guido D'Amico, Marc Kamionkowski, and Kris Sigurdson, "Dark Matter Astrophysics," in *Dark Matter and Dark Energy: A Challenge for Modern Cosmology,* ed. Sabiuno Matarrese et al. (Dordrecht, Neth.: Springer, 2011), pp. 241–72.

11. Ibid.

12. Andrea Giuliani, "Dark Matter Direct and Indirect Detection," in Matarrese et al., *Dark Matter and Dark Energy*, pp. 295–328.

13. Eric Hand, "Hunt for the Sterile Neutrino Heats Up," *Nature* 464 (2010): 334–35.

14. Robert Adler, "Neutrinos—the Next Big Small Thing," *New Scientist*, September 8, 2012, pp. 30–35; Patrick Huber and Jon Link, "Light Sterile Neutrinos: A White Paper," 2012, http://cnp.phys.vt.edu/white_paper/whitepaper.pdf (accessed September 10, 2012).

15. D'Amico, Kamionkowski, and Sigurdson, "Dark Matter Astrophysics," pp. 267–68.

16. Ibid., pp. 268–69.

17. S. Perlmutter et al., "Measurements of Omega and Lambda from 42 High-Redshift Supernovae," *Astrophysical Journal* 517 (1999): 565; A. G. Riess et al., "Observational Evidence from Supernovae for an Accelerating Universe and a Cosmological Constant," *Astronomical Journal* 116 (1998): 1009.

18. For a good review and references, see Robert P. Kirshner, "Supernovae, an Accelerating Universe and the Cosmological Constant," *Proceedings of the National Academy of Sciences* 96 (1999): 4224–27.

19. Sean M. Carroll, William H. Press, and Edwin L. Turner, "The Cosmological Constant," *Annual Reviews of Astronomy and Astrophysics* 30 (1992): 499–542.

20. Alessandra Silvestri and Mark Trodden, "Approaches to Understanding Cosmic Acceleration," *Reports on Progress in Physics* 72 (2009): 096901.

21. Steven Weinberg, "The Cosmological Constant Problem," *Reviews of Modern Physics* 61, no. 1 (1989): 1–23.

22. See Victor J. Stenger, *The Fallacy of Fine-Tuning: Why the Universe Is Not Designed for Us* (Amherst, NY: Prometheus Books, 2011), pp. 213–20.

23. Ibid., pp. 220–22.

24. William Lane Craig, "Philosophical and Scientific Pointers to Creatio Ex Nihilo," *Journal of the American Scientific Affiliation* 32, no. 1 (1980): 5–13.

25. Stephen Hawking and Roger Penrose, "The Singularities of Gravitational Collapse and Cosmology," *Proceedings of the Royal Society of London*, ser. A, 314 (1970): 529–48.

26. Victor J. Stenger, *The Comprehensible Cosmos: Where Do the Laws of Physics Come From?* (Amherst, NY: Prometheus Books, 2006), pp. 293–97.

27. Stephen Hawking, *A Brief History of Time: From the Big Bang to Black Holes* (Toronto; New York: Bantam Books, 1988), p. 50.

28. Lee Smolin, *Three Roads to Quantum Gravity* (New York: Basic Books, 2002).

29. Stenger, *Comprehensible Cosmos*, pp. 312–19.

30. Andrei Sakharov, "Vacuum Quantum Fluctuations in Curved Space," *Doklady Akademii Nauk SSSR* 177, no. 1 (1967): 70–71.

31. A. D. Linde, "Eternally Existing Self-Reproducing Chaotic Inflationary Universe," *Physics Letters B* 175, no. 4 (1986): 395–400; Andrei Linde, "The Self-Reproducing Inflationary Universe," *Scientific American Presents* (1998): 98–104; Alexander Vilenkin, *Many Worlds in One: The Search for Other Universes* (New York: Hill and Wang, 2006).

32. Victor J. Stenger, *God and the Folly of Faith: The Incompatibility of Science and Religion* (Amherst, NY: Prometheus Books, 2012), pp. 179–80.

33. William Lane Craig and James D. Sinclair, "The *Kalam* Cosmological Argument," in *The Blackwell Companion to Natural Theology*, ed. William Lane Craig and James Porter Moreland (Chichester, UK; Malden, MA: WileyBlackwell, 2009), pp. 101–201.

34. Lawrence Maxwell Krauss, *A Universe from Nothing: Why There Is Something Rather Than Nothing* (New York: Free Press, 2012).

35. David Albert, "On the Origin of Everything," *New York Times*, March 23, 2012.

36. Stenger, *Comprehensible Cosmos*.

37. Albert, "On the Origin of Everything."

38. Frank Wilczek, "The Cosmic Asymmetry between Matter and Antimatter," *Scientific American* 243, no. 6 (1980): 82–90.

39. Hawking, *Brief History of Time*; Victor J. Stenger, *Not by Design: The Origin of the Universe* (Amherst, NY: Prometheus Books, 1988).

13. SUMMARY AND CONCLUSIONS

1. Christopher Hitchens, *God Is Not Great: How Religion Poisons Everything* (New York: Twelve, 2007), p. 64.

2. Victor J. Stenger, *The Unconscious Quantum: Metaphysics in Modern Physics and Cosmology* (Amherst, NY: Prometheus Books, 1995); Victor J. Stenger, "Bioenergetic Fields," *Scientific Review of Alternative Medicine* 3, no. 1 (1997): 26–30.

3. Victor J. Stenger, *God: The Failed Hypothesis—How Science Shows That God Does Not Exist* (Amherst, NY: Prometheus Books, 2007).

4. William Grassie, *The New Sciences of Religion: Exploring Spirituality from the Outside In and Bottom Up* (New York: Palgrave Macmillan, 2010), p. 169.

5. Varadaraja V. Raman, *Truth and Tension in Science and Religion* (Center Ossipee, NH: Beech River Books, 2009), p. 115.

6. Bernard Pullman, *The Atom in the History of Human Thought: A Panoramic*

Intellectual History of a Quest That Has Engaged Scientists and Philosophers for 2,500 Years (Oxford; New York: Oxford University Press, 1998), pp. 346–47.

7. David Tong, "Is Quantum Reality Analog After All?" *Scientific American* (December 2012).

8. Stenger, *Unconscious Quantum*; Victor J. Stenger, *Quantum Gods: Creation, Chaos, and the Search for Cosmic Consciousness* (Amherst, NY: Prometheus Books, 2009).

9. For a full discussion of emergence, see Victor J. Stenger, *God and the Folly of Faith: The Incompatibility of Science and Religion* (Amherst, NY: Prometheus Books, 2012), pp. 205–18.

10. Stuart A. Kauffman, *Reinventing the Sacred: A New View of Science, Reason and Religion* (New York: Basic Books, 2008).

11. Arthur Peacocke, "The Sciences of Complexity: A New Theological Resource?" in *Information and the Nature of Reality: From Physics to Metaphysics*, ed. P. C. W. Davies and Niels Henrik Gregersen (Cambridge; New York: Cambridge University Press, 2010), pp. 249–81.

12. David Bohm and B. J. Hiley, *The Undivided Universe: An Ontological Interpretation of Quantum Theory* (London; New York: Routledge, 1993).

13. David J. Bartholomew, *God, Chance, and Purpose: Can God Have It Both Ways?* (Cambridge; New York: Cambridge University Press, 2008).

14. Stenger, *Quantum Gods*.

15. Mi-Kyong Lee, "The Distinction between Primary and Secondary Qualities in Ancient Greek Philosophy," in *Primary and Secondary Qualities: The Historical and Ongoing Debate*, ed. Lawrence Nolan (New York: Oxford University Press, 2011), pp. 15–40.

16. Stenger, *God and the Folly of Faith*, pp. 261–72.

BIBLIOGRAPHY

Adler, Robert. "Neutrinos—the Next Big Small Thing." *New Scientist*, September 8, 2012, pp. 30–35.

Albert, David. "On the Origin of Everything." *New York Times*, March 23, 2012.

al-Khalili, Jim. *The House of Wisdom: How Arabic Science Saved Ancient Knowledge and Gave Us the Renaissance*. New York: Penguin Press, 2011.

Anderson, P. W. "More Is Different." *Science* 177, no. 4047 (1972): 393–96.

ATLAS Collaboration. "Observation of a New Particle in the Search for the Standard Model Higgs Boson with the ATLAS Detector at the LHC. *Physics Letters B* 716, no. 1 (2012): 1–29.

Augustine (Bishop of Hippo), Saint, and Marcus Dods. *The City of God*. New York: Modern Library, 1993.

Bachelard, Gaston. *Les Intuitions Atomistiques [The Atomistic Intuitions]: Essai de Classification*. 2nd ed. Paris: J. Vrin, 1975.

Barbour, Ian G. *Religion and Science: Historical and Contemporary Issues*. San Francisco, CA: HarperSanFrancisco, 1997.

Barnes, V. E., P. L. Conolly, D. J. Crennell, and B. B. Culwick. "Observation of a Hyperon with Strangeness Minus Three." *Physical Review Letters* 12, no. 8 (1964): 204–206.

Bartholomew, David J. *God, Chance, and Purpose: Can God Have It Both Ways?* Cambridge; New York: Cambridge University Press, 2008.

Bartusiak, Marcia. *The Day We Found the Universe*. New York: Pantheon Books, 2009.

Basil, Robert, ed. *Not Necessarily the New Age: Critical Essays*. Amherst, NY: Prometheus Books, 1988.

Berginger, J., et al. "The Review of Particle Physics." *Physical Review* D86 (2012): 010001.

Bethe, Hans A. "The Electromagnetic Shift of Energy Levels." *Physical Review* 72, no. 4 (1947): 339–41.

Blom, Philipp. *A Wicked Company: The Forgotten Radicalism of the European Enlightenment*. New York: Basic Books, 2010.

Bohm, David, and B. J. Hiley. *The Undivided Universe: An Ontological Interpretation of Quantum Theory*. London; New York: Routledge, 1993.

Brewton-Parker College. "History of the Development of the Periodic Table of Elements." April 15, 2010. http://www.bpc.edu/mathscience/chemistry/history_of_the_periodic_table.html.

Brown, Alison. *The Return of Lucretius to Renaissance Florence.* Cambridge, MA: Harvard University Press, 2010.

Brundell, Barry. *Pierre Gassendi: From Aristotelianism to a New Natural Philosophy.* Dordrecht, Neth.; Boston; Norwell, MA: D. Reidel, 1987.

Byers, Nina. "E. Noether's Discovery of the Deep Connection between Symmetries and Conservation Laws." Paper presented at the Israel Mathematical Conference, Bar Ilan University, Tel Aviv, Israel, December 2–3, 1996.

Byrne, Alex, and David R. Hilbert. "Are Colors Secondary?" In *Primary and Secondary Qualities: The Historical and Ongoing Debate,* pp. 339–61. Edited by Lawrence Nolan. New York: Oxford University Press, 2011.

Capra, Fritjof. *The Tao of Physics: An Exploration of the Parallels between Modern Physics and Eastern Mysticism.* Berkeley, CA: Shambhala, 1975.

Carrier, Richard. "Christianity Was Not Responsible for Modern Science." In *The Christian Delusion: Why Faith Fails,* pp. 396–419. Edited by John W. Loftus. Amherst, NY: Prometheus Books, 2010.

———. "Predicting Modern Science: Epicurus vs. Mohammed." Secular Web. June 22, 2004. http://www.infidels.org/kiosk/article362.html.

Carroll, Sean M., William H. Press, and Edwin L. Turner. "The Cosmological Constant." *Annual Reviews of Astronomy and Astrophysics* 30 (1992): 499–542.

Chew, Geoffrey F. "S-Matrix Theory of Strong Interactions without Elementary Particles." *Reviews of Modern Physics* 34, no. 3 (1962): 394–401.

Clausius, Rudolf. "Ueber die bewegende Kraft der Wärme und die Gesetze, welche sich daraus für die Wärmelehre selbst ableiten lassen." *Annalen der Physik und Chemie* 155, no. 3 (1850): 368–94.

Clayton, Philip, and P. C. W. Davies, eds. *The Re-Emergence of Emergence: The Emergentist Hypothesis from Science to Religion.* Oxford; New York: Oxford University Press, 2006.

CMS Collaboration. "Observation of a New Boson at a Mass of 125 Gev with the CMS Experiment at the LHC." *Physics Letters B* 716, no. 1 (2012): 30–61.

Craig, William Lane. "Philosophical and Scientific Pointers to Creatio Ex Nihilo." *Journal of the American Scientific Affiliation* 32, no. 1 (1980): 5–13.

Craig, William Lane, and James D. Sinclair. "The *Kalam* Cosmological Argument." In *The Blackwell Companion to Natural Theology,* pp. 101–201. Edited by William Lane Craig and James Porter Moreland. Chichester, UK; Malden, MA: Wiley-Blackwell, 2009.

Craig, William Lane, and James Porter Moreland, eds. *The Blackwell Companion to Natural Theology.* Chichester, UK; Malden, MA: Wiley-Blackwell, 2009.

Dalton, John. *A New System of Chemical Philosophy*. Manchester; London, 1808.

D'Amico, Guido, Marc Kamionkowski, and Kris Sigurdson. "Dark Matter Astrophysics." In *Dark Matter and Dark Energy: A Challenge for Modern Cosmology*, pp. 241–72. Edited by Sabiuno Matarrese et al. Dordrecht, Neth.: Springer, 2011.

Davies, C. T. H., E. Follana, A. Gray, and G. P. Lepage. "High-Precision Lattice QCD Confronts Experiment." *Physical Review Letters* 92, no. 2 (2004): 022001–022005.

de la Mettrie, Julien O. *L'Homme Machine*. Lyde, 1748.

Dirac, P. A. M. *The Principles of Quantum Mechanics*. 4th ed. Oxford: Clarendon Press, 1958. First published 1930.

Dolgert, Drew. "Einstein's Explanation of Brownian Motion." Fowler's Physics Applets. September 22, 1998. http://Galileo.phys.Virginia.EDU/classes/109N/more_stuff/Applets/brownian/applet.html.

Duhem, Pierre Maurice Marie. *The Aim and Structure of Physical Theory*. Princeton, NJ: Princeton University Press, 1954.

———. *Thermodynamics and Chemistry: A Non-Mathematical Treatise for Chemists and Students of Chemistry*. New York; London: J. Wiley and Sons; Chapman and Hall, 1903.

———. *To Save the Phenomena: An Essay on the Idea of Physical Theory From Plato to Galileo*. Chicago: University of Chicago Press, 1969.

Dyson, Freeman J. "The Radiation Theories of Tomonaga, Schwinger, and Feynman." *Physical Review* 75, no. 3 (1949): 486–502.

Einstein, Albert, R. Fürth, and Alfred Denys Cowper. *Investigations on the Theory of the Brownian Movement*. London: Methuen, 1926.

Englert, F., and R. Brout. "Broken Symmetry and the Mass of Gauge Vector Bosons." *Physical Review Letters* 13 (1964): 321.

Faraday, Michael. "On the Physical Lines of Magnetic Force." In *Great Books of the Western World. Encyclopaedia Britannica, Inc., in Collaboration with the University of Chicago*, p. 530. Vol. 45. Edited by Robert Maynard Hutchins. Chicago: W. Benton, 1952.

Ferguson, Marilyn. *The Aquarian Conspiracy: Personal and Social Transformation in the 1980s*. Los Angeles; New York: J. P. Tarcher, 1980.

Feynman, Richard P. *The Genius of Science: A Portrait Gallery*. Oxford; New York: Oxford University Press, 2000.

———. *QED: The Strange Theory of Light and Matter*. Princeton, NJ: Princeton University Press, 1985.

———. "Space-Time Approach to Quantum Electrodynamics." *Physical Review* 76, no. 6 (1949): 769–89.

———. "The Theory of Positrons." *Physical Review* 76 (1949): 749–59.

Freund, Ida. 1904. *The Study of Chemical Composition. An Account of Its Method and*

Historical Development, with Illustrative Quotations. Cambridge: Cambridge University Press, 1904.

Gallagher, Mary. "Dryden's Translation of Lucretius." *Huntington Library Quarterly* 28, no. 1 (1964): 19–29.

Georgi, Howard, and Sheldon Glashow. "Unity of All Elementary Particle Forces." *Physical Review Letters* 32 (1974): 438–41.

Giuliani, Andrea. "Dark Matter Direct and Indirect Detection." In *Dark Matter and Dark Energy: A Challenge for Modern Cosmology*, pp. 295–328. Edited by Sabiuno Matarrese, Monica Colpi, Vittorio Gorini, and Ugo Moschella. Dordrecht, Neth.: Springer, 2011.

Goldsmith, Donald. *Einstein's Greatest Blunder? The Cosmological Constant and Other Fudge Factors in the Physics of the Universe*. Cambridge, MA: Harvard University Press, 1995.

Grassie, William. *The New Sciences of Religion: Exploring Spirituality from the Outside In and Bottom Up*. New York: Palgrave Macmillan, 2010.

Greenblatt, Stephen. *The Swerve: How the World Became Modern*. New York: W. W. Norton, 2011.

Guralnik, G. G., C. R. Hagen, and T. W. Kibble. "Global Conservation Laws and Massless Particles." *Physical Review Letters* 13 (1964): 585.

Guth, Alan H. "Inflationary Universe: A Possible Solution to the Horizon and Flatness Problems." *Physical Review D* 23, no. 2 (1981): 347–56.

———. *The Inflationary Universe: The Quest for a New Theory of Cosmic Origins*. Reading, MS: Addison-Wesley, 1997.

Hannam, James. *God's Philosophers: How the Medieval World Laid the Foundations of Modern Science*. London: Icon, 2009.

Hargraves, Robert. "Aim High: Using Thorium to Address Environmental Problems." YouTube video, 59:50. From Google Tech Talk. Posted by "GoogleTechTalks." May 26, 2009. http://www.youtube.com/watch?v=VgKfS74hVvQ.

Hargraves, Robert, and Ralph Moir. "Liquid Fluoride Thorium Reactors: An Old Idea in Nuclear Power Gets Reexamined." *American Scientist* 98, no. 4 (2010): 304–13.

———. "Liquid Fuel Nuclear Reactors." *Physics & Society* 40, no. 1 (2011): 6–10.

Harman, P. M. *Energy, Force, and Matter: The Conceptual Development of Nineteenth-Century Physics*. Cambridge; New York: Cambridge University Press, 1982.

Hawking, Stephen. *A Brief History of Time: From the Big Bang to Black Holes*. Toronto; New York: Bantam Books, 1988.

Hawking, Stephen, and Roger Penrose. "The Singularities of Gravitational Collapse and Cosmology." *Proceedings of the Royal Society of London*, ser. A, 314 (1970): 529–48.

Heisenberg, Werner. *Der Teil und das Ganze*. Munich: R. Piper, 1969.

Hewlett, Richard G., and Jack M. Holl. *Atoms for Peace and War, 1953–1961: Eisenhower and the Atomic Energy Commission*. Berkeley: University of California Press, 1989.

Higgs, Peter W. "Broken Symmetries and the Masses of Gauge Bosons." *Physical Review Letters* 13 (1964): 508.

Hilbert, David. "On the Infinite." In *Philosophy of Mathematics*. Edited by Paul Benacerraf and Hillary Putnam. Englewood Cliffs, NJ: Prentice-Hall, 1964.

Hitchens, Christopher. *God Is Not Great: How Religion Poisons Everything*. New York: Twelve, 2007.

Hoddeson, Lillian. *The Rise of the Standard Model: Particle Physics in the 1960s and 1970s*. New York: Cambridge University Press, 1997.

Holden, Norman E. "History of the Origin of the Chemical Elements and Their Discoverers." National Nuclear Data Center. 2004. http://www.nndc.bnl.gov/content/elements.html.

Holt, Jim. "Physicists, Stop the Churlishness." *New York Times*, June 10, 2012.

Holton, Gerald. *Victory and Vexation in Science: Einstein, Bohr, Heisenberg, and Others*. Cambridge, MA: Harvard University Press, 2005.

Huber, Patrick, and Jon Link. "Light Sterile Neutrinos: A White Paper." 2012.

Hutchins, Robert Maynard, ed. *Great Books of the Western World. Encyclopaedia Britannica, Inc., in Collaboration with the University of Chicago*. Vol. 45. Chicago: W. Benton, 1952.

International Atomic Energy Agency, World Health Organization, and United Nations Development Agency. "Chernobyl: The True Scale of the Accident." Press release, September 5, 2005.

Iocco, F., G. Mangano, G. Miele, O. Pisanti, and P. D. Serpico. "Primordial Nucleosynthesis: From Precision Cosmology to Fundamental Physics." *Physics Reports* 472 (2009): 1–76.

Jarosik, N., C. L. Bennett, J. Dunkley, and B. Gold. "Seven-Year Wilkinson Microwave Anisotropy Probe (WMAP). Observations: Sky Maps, Systematic Errors, and Basic Results." *Astrophysical Journal* 192 (2011): 14.

Joule, James Prescott. "On the Mechanical Equivalent of Heat." *Philosophical Transactions of the Royal Society of London* 140, no. 1 (1850): 61–82.

———. "On the Rarefaction and Condensation of Air." *Philosophical Magazine, Scientific Papers* 172 (1845).

Joy, Lynn Sumida. *Gassendi, the Atomist: Advocate of History in an Age of Science*. Cambridge; New York: Cambridge University Press, 1987.

Kaiser, David. "Physics and Feynman's Diagrams." *American Scientist* 93 (2005): 156–65.

Kang, Jungmin, and Frank N. von Hippel. "U-232 and the Proliferation-Resistance of U-233 in Spent Fuel." *Science & Global Security* 9 (2001): 1–32.

Kauffman, Stuart A. *Reinventing the Sacred: A New View of Science, Reason and Religion.* New York: Basic Books, 2008.

Kazanas, Demos. "Dynamics of the Universe and Spontaneous Symmetry Breaking." *Astrophysical Journal* 241 (1980): L59–L63.

Kirk, Geoffrey S., John E. Raven, and Malcolm Schofield. *The Presocratic Philosophers: A Critical History with a Selection of Texts.* 2nd ed. Cambridge; New York: Cambridge University Press, 1983.

Kirsch, Jonathan. *God against the Gods: The History of the War between Monotheism and Polytheism.* New York: Viking Compass, 2004.

Kirshner, Robert P. "Supernovae, an Accelerating Universe and the Cosmological Constant." *Proceedings of the National Academy of Sciences* 96 (1999): 4224–27.

Koyré, Alexandre. *Metaphysics and Measurement.* Yverdon, Switz.; Langhorne, PA: Gordon and Breach Science, 1992.

Krauss, Lawrence Maxwell. *A Universe from Nothing: Why There Is Something Rather Than Nothing.* New York: Free Press, 2012.

Kuhn, Thomas. "Energy Conservation as an Example of Simultaneous Discovery." In *Critical Problems in the History of Science,* pp. 321–56. Edited by Marshall Clagett. Madison: University of Wisconsin Press, 1969.

Lamb, Willis E., Jr., and Robert C. Retherford. "Fine Structure of the Hydrogen Atom by a Microwave Method." *Physical Review* 72, no. 3 (1947): 241–43.

Lederman, Leon M., and Christopher T. Hill. *Symmetry and the Beautiful Universe.* Amherst, NY: Prometheus Books, 2004.

Lederman, Leon M., and Dick Teresi. *The God Particle: If the Universe Is the Answer, What Is the Question?* Boston: Houghton Mifflin, 2006.

Lee, Mi-Kyong. "The Distinction between Primary and Secondary Qualities in Ancient Greek Philosophy." In *Primary and Secondary Qualities: The Historical and Ongoing Debate,* pp. 15–40. Edited by Lawrence Nolan. New York: Oxford University Press, 2011.

Lehninger, Albert L. *Bioenergetics: The Molecular Basis of Biological Energy Transformations.* 2nd ed. Menlo Park, CA: W. A. Benjamin, 1971.

Lemaître, Georges. "Un univers homogène de masse constante et de rayon croissant rendant compte de la vitesse radiale des nébuleuses extra-galactiques" [A Homogeneous Universe of Constant Mass and Growing Radius Accounting for the Radial Velocity of Extragalactic Nebulae]. *Annales de la Société Scientifique de Bruxelles* 47 (1927): 49.

Lindberg, David C. *The Beginnings of Western Science: The European Scientific Tradition in Philosophical, Religious, and Institutional Context, Prehistory to A.D. 1450.* 2nd ed. Chicago: University of Chicago Press, 2007.

Linde, A. D. "Eternally Existing Self-Reproducing Chaotic Inflationary Universe." *Physics Letters B* 175, no. 4 (1986): 395–400.

———. "A New Inflationary Universe Scenario: A Possible Solution of the Horizon, Flatness, Homogeneity, Isotropy and Primordial Monopole Problems." *Physics Letters B* 108 (1982): 389.

———. "The Self-Reproducing Inflationary Universe." *Scientific American Presents* (1998): 98–104.

Lindley, David. *Boltzmann's Atom: The Great Debate That Launched a Revolution in Physics*. New York: Free Press, 2001.

LoLordo, Antonia. *Pierre Gassendi and the Birth of Early Modern Philosophy*. New York: Cambridge University Press, 2007.

Mackay, Alan L. *A Dictionary of Scientific Quotations*. Bristol; Philadelphia: A. Hilger, 1991.

Maimonides, Moses, Julius Guttmann, Chaim Rabin, and Daniel H. Frank. *The Guide of the Perplexed*. Indianapolis, IN: Hackett, 1995.

Mandl, F., and G. Shaw. *Quantum Field Theory*. Rev. ed.. Chichester, UK; New York: Wiley, 1993.

Martin, Richard. *Superfuel: Thorium, the Green Energy Source for the Future*. New York: Palgrave Macmillan, 2012.

Matarrese, Sabiuno, Monica Colpi, Vittorio Gorini, and Ugo Moschella, eds. *Dark Matter and Dark Energy: A Challenge for Modern Cosmology*. Dordrecht, Neth.: Springer, 2011.

Maund, Barry. "Color Eliminativism." In *Primary and Secondary Qualities: The Historical and Ongoing Debate*, pp. 362–85. Edited by Lawrence Nolan. New York: Oxford University Press, 2011.

Nafe, J. E., E. B. Nelson, and I. I. Rabi. 1947. "Hyperfine Structure of Atomic Hydrogen and Deuterium." *Physical Review* 71, no. 12 (1947): 914–15.

Newton, Isaac. *Mathematical Principles of Natural Philosophy*. New York: Greenwood Press, 1969.

Nishino, H., S. Clark, K. Abe, and Y. Hayato. "Search for Proton Decay via $p \rightarrow e^+ + \pi^0$ and $p \rightarrow \mu^+ + \pi^0$ in a Large Water Cherenkov Detector." *Physical Review Letters* 102 (2009): 141801.

Niven, William Davidson, ed. *The Scientific Papers of James Clerk Maxwell*. Vol 2. Cambridge: Cambridge University Press, 1890.

Nolan, Lawrence. Introduction to *Primary and Secondary Qualities: The Historical and Ongoing Debate*. Edited by Lawrence Nolan. New York: Oxford University Press, 2011.

———, ed. *Primary and Secondary Qualities: The Historical and Ongoing Debate*. New York: Oxford University Press, 2011.

Nuclear Energy Institute. "Nuclear Energy around the World." 2012. http://www
.nei.org/resourcesandstats/nuclear_statistics/worldstatistics/.

O'Connor, Eugene. *The Essential Epicurus: Letters, Principal Doctrines, Vatican Sayings, and Fragments.* Amherst, NY: Prometheus Books, 1993.

O'Keefe, Tim. "Epicurus (341–271 BCE)." *Internet Encyclopedia of Philosophy.* 2005. http://www.iep.htm.edu/epicur/#SSH3c.ii.

Peacocke, Arthur. *The Sciences of Complexity: A New Theological Resource?* In *Information and the Nature of Reality: From Physics to Metaphysics,* pp. 249–81. Edited by P. C. W. Davies and Niels Henrik Gregersen. Cambridge; New York: Cambridge University Press, 2010.

Perlmutter, S., et al. "Measurements of Omega and Lambda from 42 High-Redshift Supernovae." *Astrophysical Journal* 517 (1999): 565.

Perrin, Jean. *Brownian Movement and Molecular Reality.* Mineola, NY: Dover, 2005.

Pullman, Bernard. *The Atom in the History of Human Thought; a Panoramic Intellectual History of a Quest That Has Engaged Scientists and Philosophers for 2,500 Years.* Oxford; New York: Oxford University Press, 1998.

Pyle, Andrew. *Atomism and Its Critics: Problem Areas Associated with the Development of the Atomic Theory of Matter from Democritus to Newton.* Bristol, UK: Thoemmes Press, 1995.

Raman, Varadaraja V. *Truth and Tension in Science and Religion.* Center Ossipee, NH: Beech River Books, 2009.

Reiss, John O. *Not by Design: Retiring Darwin's Watchmaker.* Berkeley: University of California Press, 2009.

Riess, A. G., A. V. Filippenko, P. Challis, A. Clocchiatti, A. Diercks, P. M. Garnavich, R. L. Gilliland, C. J. Hogan, S. Jha, and R. P. Kirshner. "Observational Evidence from Supernovae for an Accelerating Universe and a Cosmological Constant." *Astronomical Journal* 116 (1998): 1009.

Rosenberg, Alex. *Darwinian Reductionism; or, How to Stop Worrying and Love Molecular Biology.* Chicago; London: University of Chicago Press, 2006.

Rubin, Julian T. "James Prescott Joule: The Discovery of the Mechanical Equivalent of Heat." Following the Path of Discovery. 2011. http://www.juliantrubin .com/bigten/mechanical_equivalent_of_heat.html.

Sakharov, Andrei. "Vacuum Quantum Fluctuations in Curved Space." *Doklady Akademii Nauk SSSR* 177, no. 1 (1967): 70–71.

Sample, Ian. *Massive: The Missing Particle That Sparked the Greatest Hunt in Science.* New York: Basic Books, 2012.

Schweber, Silvan S. *QED and the Men Who Made It: Dyson, Feynman, Schwinger, and Tomonaga.* Princeton, NJ: Princeton University Press, 1994.

Schwinger, Julian. "On Quantum-Electrodynamics and the Magnetic Moment of the Electron." *Physical Review* 73, no. 4 (1948): 416–17.

Sedley, D. N. *"Creationism and Its Critics in Antiquity."* Berkeley: University of California Press, 2007.

Shannon, Claude Elwood, and Warren Weaver. *The Mathematical Theory of Communication*. Urbana: University of Illinois Press, 1949.

Shiltsev, Vladimir. "Mikhail Lomonosov and the Dawn of Russian Science." *Physics Today* 65, no. 2 (2012): 40–45.

Silvestri, Alessandra, and Mark Trodden. "Approaches to Understanding Cosmic Acceleration." *Reports on Progress in Physics* 72 (2009): 096901.

Singer, Charles, ed. *Studies in the History and Methods of Science*. Oxford: Oxford University Press, 1917.

Sinha, Sukanya, and Rafael D. Sorkin. "A Sum-over-Histories Account of an EPR(B) Experiment." *Foundations of Physics Letters* 4, no. 4 (1991): 303–35.

Smith, Timothy Paul. *Hidden Worlds: Hunting for Quarks in Ordinary Matter*. Princeton, NJ: Princeton University Press, 2003.

Smolin, Lee. *The Life of the Cosmos*. New York: Oxford University Press, 1997.

———. "A Perspective on the Landscape Problem." Perimeter Institute for Theoretical Physics. February 16, 2012. http://arxiv.org/pdf/1202.3373v1.pdf.

———. "The Status of Cosmological Natural Selection." Perimeter Institute for Theoretical Physics. February 2, 2008. http://arxiv.org/pdf/hep-th/0612185v1.pdf.

———. *Three Roads to Quantum Gravity*. New York: Basic Books, 2002.

Smoot, George, and Keay Davidson. *Wrinkles in Time: Witness to the Birth of the Universe*. New York: Harper Perennial, 2007.

Stallings, A. E., and Richard Jenkyns. *Lucretius: The Nature of Things*. London; New York: Penguin, 2007.

Steffens, Henry John. *James Prescott Joule and the Concept of Energy*. Folkestone, UK; New York: Dawson Science History, 1979.

Stenger, Victor J. "Bioenergetic Fields." *Scientific Review of Alternative Medicine* 3, no. 1 (1997): 26–30.

———. *The Comprehensible Cosmos: Where Do the Laws of Physics Come From?* Amherst, NY: Prometheus Books, 2006.

———. *The Fallacy of Fine-Tuning: Why the Universe Is Not Designed for Us*. Amherst, NY: Prometheus Books, 2011.

———. *God and the Folly of Faith: The Incompatibility of Science and Religion*. Amherst, NY: Prometheus Books, 2012.

———. *Not by Design: The Origin of the Universe*. Amherst, NY: Prometheus Books, 1988.

———. *Quantum Gods: Creation, Chaos, and the Search for Cosmic Consciousness*. Amherst, NY: Prometheus Books, 2009.

———. *Timeless Reality: Symmetry, Simplicity, and Multiple Universes*. Amherst, NY: Prometheus Books, 2000.

———. *The Unconscious Quantum: Metaphysics in Modern Physics and Cosmology*. Amherst, NY: Prometheus Books, 1995.

Stückelberg, Ernst. "La mécanique du point matériel en théorie de relativité et en théorie des quanta." *Helvetica Physica Acta* 15 (1942): 23–37.

Super-Kamiokande Collaboration. "Evidence for Oscillation of Atmospheric Neutrinos." *Physical Review Letters* 81 (1998): 1562–67.

Sutherland, Roderick I. "Bell's Theorem and Backward-in-Time Causality." *International Journal of Theoretical Physics* 22, no. 4 (1883): 377–84.

Svoboda, R., and K. Gordan. "Neutrinos in the Sun." Astronomy Picture of the Day. June 5, 1998. http://apod.nasa.gov/apod/ap980605.html.

Sylla, Edith Dudley. *The Oxford Calculators and the Mathematics of Motion, 1320–1350: Physics and Measurement by Latitudes*. New York: Garland, 1991.

Thomson, William. "On an Absolute Temperature Scale Founded on Carnot's Theory of the Motive Power of Heat, and Calculated from Regnault's Observations." *Philosophical Magazine* 1 (1848).

———. *Mathematical and Physical Papers*. Cambridge: Cambridge University Press, 1882.

Tomonaga, Sin-Itiro. "On a Relativistically Invariant Formulation of the Quantum Theory of Wave Fields." *Progress in Theoretical Physics* 1, no. 2 (1946): 27–42.

Tong, David. "Is Quantum Reality Analog After All?" *Scientific American* (December 2012).

Tsujikawa, Shinji. "Dark Energy: Investigation and Modeling." In *Dark Matter and Dark Energy: A Challenge for Modern Cosmology*. Edited by Sabiuno Matarrese, Monica Colpi, Vittorio Gorini, and Ugo Moschella. Dordrecht, Neth.: Springer, 2011.

Turnbull, H., ed. *The Correspondence of Isaac Newton*. Vol. 1. Cambridge: Cambridge University Press, 1959.

Tyndall, John, and William Francis. *Scientific Memoirs, Selected from the Transactions of Foreign Academies of Science, and from Foreign Journals. Natural Philosophy*. London: Taylor and Francis, 1853.

Vilenkin, Alexander. *Many Worlds in One: The Search for Other Universes*. New York: Hill and Wang, 2006.

Wallace, Wes. "The Vibrating Nerve Impulse in Newton, Willis and Gassendi: First Steps in a Mechanical Theory of Communication." *Brain and Cognition* 51 (2003): 66–94.

Weinberg, Alvin Martin. *The First Nuclear Era: The Life and Times of a Technological Fixer*. New York: AIP Press, 1994.

Weinberg, Steven. "The Cosmological Constant Problem." *Reviews of Modern Physics* 61, no. 1 (1989): 1–23.

———. *Dreams of a Final Theory*. New York: Pantheon Books, 1992.

———. *The First Three Minutes: A Modern View of the Origin of the Universe*. New York: Basic Books, 1977.

White, Michael. *Isaac Newton: The Last Sorcerer*. Reading, MA: Addison-Wesley, 1997.

Wilczek, Frank. "The Cosmic Asymmetry between Matter and Antimatter." *Scientific American* 243, no. 6 (1980): 82–90.

Wilson, Catherine. *Epicureanism at the Origins of Modernity*. Oxford; New York: Clarendon Press; Oxford University Press, 2008.

World Health Organization. "Air Quality and Health: Fact Sheet no. 313."

World Nuclear Association. "Nuclear Power Reactors." 2011.

Yergin, Daniel. *The Quest: Energy, Security and the Remaking of the Modern World*. New York: Penguin Press, 2011.

Yukawa, Hideki. "On the Interaction of Elementary Particles." *Progress in Theoretical Physics* 17 (1935): 48–56.

ABOUT THE AUTHOR

Victor J. Stenger grew up in a Catholic working-class neighborhood in Bayonne, New Jersey. His father was a Lithuanian immigrant; his mother, the daughter of Hungarian immigrants. He attended public schools and received a bachelor of science degree in electrical engineering from Newark College of Engineering (now New Jersey Institute of Technology) in 1956. While at NCE, he was editor of the student newspaper and received several journalism awards.

Moving to Los Angeles on a Hughes Aircraft Company fellowship, Dr. Stenger received a master of science degree in physics from UCLA in 1959 and a doctorate in physics in 1963. He then took a position on the faculty of the University of Hawaii and retired to Colorado in 2000. He currently is adjunct professor of philosophy at the University of Colorado and emeritus professor of physics at the University of Hawaii. Dr. Stenger has also held visiting positions on the faculties of the University of Heidelberg in Germany and the University of Oxford in England, and he has been a visiting researcher at Rutherford Laboratory in England, the National Nuclear Physics Laboratory in Frascati, Italy, and the University of Florence in Italy.

His research career spanned the period of great progress in elementary particle physics that ultimately led to the current *standard model*. He participated in experiments that helped establish the properties of strange particles, quarks, gluons, and neutrinos. He also helped pioneer the emerging fields of very high-energy gamma ray and neutrino astronomy. In his last project before retiring, Dr.

Stenger collaborated on the underground experiment in Japan that in 1998 showed for the first time that the neutrino has mass. The Japanese leader of this project, Masatoshi Koshiba, shared the 2002 Nobel Prize in Physics for this work.

Victor J. Stenger has had a parallel career as an author of critically well-received, popular-level books that interface between physics and cosmology and philosophy, religion, and pseudoscience. His 2007 book, *God: The Failed Hypothesis,* made the *New York Times* bestseller list in March of that year.

Dr. Stenger and his wife, Phylliss, have been happily married since 1962 and have two children and four grandchildren. They celebrated their golden wedding anniversary on October 6, 2012. They now live in Lafayette, Colorado, and travel the world as often as they can.

Dr. Stenger maintains a popular website where much of his writing can be found, at http://www.colorado.edu/philosophy/vstenger/. He also maintains an e-mail discussion list, avoid-L, where the topics range from his own writings to the whole gamut of intellectual discourse and politics.

OTHER BOOKS BY VICTOR J. STENGER

Not by Design: The Origin of the Universe (1988)

Physics and Psychics: The Search for a World beyond the Senses (1990)

The Unconscious Quantum: Metaphysics in Modern Physics and Cosmology (1995)

Timeless Reality: Symmetry, Simplicity, and Multiple Universes (2000)

Has Science Found God? The Latest Results in the Search for Purpose in the Universe (2003)

The Comprehensible Cosmos: Where Do the Laws of Physics Come From? (2006)

God: The Failed Hypothesis—How Science Shows That God Does Not Exist (2007)

Quantum Gods: Creation, Chaos, and the Search for Cosmic Consciousness (2009)

The New Atheism: Taking a Stand for Science and Reason (2009)

The Fallacy of Fine-Tuning: Why the Universe Is Not Designed for Us (2011)

God and the Folly of Faith: The Incompatibility of Science and Religion (2012)

INDEX

absolute zero, 99

acupuncture, 263

Adelard of Bath, 49

Aeneid (Virgil), 52

aether, 123, 126, 131, 133, 267, 270

Agamemnon, 37

age of reason, 87

air thermometer, 99

Albert, David, 257, 258

Albert the Great, 80

alchemy, 24, 79, 80, 82, 124

    in nuclear physics, 151

Alhacen (Ibn al-Haytham), 120

alpha radiation, 150, 169

American Physical Society, 232

Ampere's law, 131

Anderson, Carl D., 158, 186, 198

Anderson, Phillip, 232

Annalen der Physik (journal), 96

ant colony, 270

anthropic coincidences, 146

anthropic principle, 146

antielectrons, 158

antimatter in the universe, 173

antineutrons, 158

    discovery of, 214

antiprotons, 158

    discovery of, 214

Aquinas, St. Thomas, 22, 80

Arabic atomism, 50

Archimedes of Syracuse, 34

Aristarchus of Samos, 34, 64

Aristotle, 11, 13, 22, 30, 33, 40, 42, 43, 47, 49, 74, 88, 120, 123, 262, 271, 272

    physics of, 72

Arrhenius, Svante, 113

arrow of time, 110, 207

astrology, 57

asymptotic freedom, 225

ATLAS (A Toroidal LHC Apparatus), 233

atomic bomb, 174

Atomic Energy Commission, 177

atoms, properties of,

    atomic mass, 162

    atomic number, 162

    atomic shells, 164

    atomic spectra, 146

    atomic subshells, 164

    atomic weight, 85, 162, 168

Augustine of Hippo, 22, 45, 47, 48, 62, 253

Aurelius, Marcus, 44

Averroes (Ibn Rushd), 51

317

Avogadro, Lorenzo Romano Amedeo Carlo, 102
  Avogadro's law, 87, 102
  Avogadro's number, 102, 103, 104, 118
axion, 250
Ayer, A. J., 114
azimuthal quantum number, 153, 156

Bachelard, Gaston, 40
Bacon, Francis, 121
baryon, 204
  baryon density of the universe, 248
  baryon number, 205
  baryon number conservation, 204, 206, 255
Becquerel, Henri, 150
Becquerel rays, 115
Bentley, Richard, 75
Bernoulli, Daniel, 71, 75, 91, 92
Berthelot, Marcellin, 87
beta decay, 186, 150,169, 218, 200
Bethe, Hans, 189, 190, 197
Bevatron, 204, 214
big bang, 253, 256, 274
bioenergetic field, 263
biomass, 175
black body radiation, 146
black body spectrum, 148
Bohm, David, 159
Bohmian quantum mechanics, 272

Bohr, Niels, 145, 155
  Bohr hypothesis, 152
  Bohr's model of the hydrogen atom, 152, 153
Bohr-Sommerfeld model of the atom, 154, 156, 157
Boltzmann, Ludwig, 101, 104, 105, 110, 112, 113, 114, 115, 116, 148, 272
  Boltzmann's constant, 102, 106, 107
Bonfire of the Vanities, 54
Book of Revelation, 82
bootstrap theory, 208–10
Born, Max, 155, 159
boson defined, 157
bottom-up causality, 271
Boyle, Robert, 59, 72, 75, 82, 91
  Boyle's law, 75, 82, 91, 101
Bracciolini, Gian Francesco Poggio, 51
Bradwardine, Thomas, 63
Brahe, Tycho, 67
branes, 17
breeder reactors, 175
Brentano, Franz, 116
broken symmetry/symmetries, 204, 206
Brookhaven National Laboratory, 83, 203, 214
Brout, Robert, 226
Brown, Robert, 117
  Brownian motion, 40, 117, 148

Bruno, Giordano, 56, 72
bubble chamber, 204, 214
buckyballs, 154
Buddhism, atoms in, 29
Buridan, Jean, 63
Bush, George H. W., 231

Caesar, Julius, 12
caloric theory of heat, 96
Capra, Fritjof, 210
carbon sequestration, 183
Carnap, Rudolf, 114
Carnot, Nicolas Léonard Sadi,
    93, 97, 99
    Carnot cycle or Carnot en-
        gine, 93, 94
Carrier, Richard, 33, 41
Cartesian coordinate system,
    130, 141
Cartesian model, 59
central limit theorem, 107
CERN, 203, 215, 223, 232, 234
Cesium-133, 137
Chadwick James, 151, 158, 186
chain reaction, 174
charge conservation, 168, 206,
    220
charge-conjugation, 207
    charge-conjugation symme-
        try, 142
Charles law, 101
chemical atom
    definition of, 150
    model of, 165

chemical elements, 162
chemical periodic table, 264
Chemical Society, 87
Chernobyl, 177, 178, 182
Chew, Geoffrey, 208
chi, 263
Chinese medicine, 263
Church of England, 55
Cicero, 44, 45, 88
Clapeyron, Benoît Paul Émile,
    97, 101
Clausius, Rudolf, 96, 100, 101,
    104
Clinton, Bill, 232
clock paradox, 135
CMS (CERN experiment), 233
color charge, 224, 225, 226
Columbia University, 187, 188
complementary and alternative
    medicine, 263
complex number, 155
Comte, Auguste, 113
conservation
    of angular momentum, 143
    of energy, 84, 96, 97, 100,
        143, 170
    of linear momentum, 143
    of momentum, 69
Constantine (Roman emperor),
    62
Copernicus, Nicolaus, 64
Copernican model, 56, 57, 64
corpuscular nature of light
    (Newton), 122

cosmic acceleration, 251

Cosmic Background Explorer (COBE), 245

cosmic background radiation, 245

cosmic rays, 264

cosmological constant, 192, 243, 251, 252

    cosmological-constant problem, 192

Cossa, Baldassare. *See* Pope John XXIII

CP symmetry, 207, 255

CPT symmetry, 207, 208

Craig, William Lane, 254, 256

creation ex nihilo, 253

critical density of universe, 244

Cronin, James, 207

Ctesibius of Alexandria, 34

Curie, Marie, 150

Curie, Pierre, 150

Dalton, John, 79, 84, 85

Dante Alighieri, 31

Dark Ages, 13, 50, 62, 64, 81

dark energy, 236, 249, 251, 255, 263

dark matter, 236, 247, 248, 249, 251, 255, 263

Darwin, Charles, 68

Darwinian evolution, 88, 273

Davisson, Clinton, 154

Dawkins, Richard, 257

de Broglie, Louis-Victor-Pierre-Raymond, 154, 159, 169

de Broglie wavelength, 154, 155, 160

Dee, John, 80

deism, 87, 273

Democritus, 11, 23, 25, 26, 27, 29, 33, 38, 51, 73–75, 242, 261, 266

Democritus-Epicurus view of the soul, 50

De rerum natura [*The Nature of Things*] (Lucretius), 13, 31, 35, 37, 51, 52, 54, 55, 72

Descartes, René, 56, 58, 72, 77, 124

Deville, Henri Sainte-Claire, 87

d'Holbach, Paul-Henry Thiry, Baron, 76

Diderot, Denis, 76, 87

diffraction, 122, 160

Diogenes Laertius, 24, 31, 58

Dirac, Paul, 156, 158, 159, 187, 188

    Dirac equation, 158, 187

    Dirac field, 269

    quantum mechanics of, 156–59

    theory of the electron of, 158, 187

DNA, 88, 270

Doppler effect, 132

Dryden, John, 35, 36

duality of mind and body, 267

Duhem, Pierre, 62, 64, 88, 110

Dumas, Jean-Baptiste, 87, 88

Dyson, Freeman, 189

Eastern mysticism, 210
Eddington, Arthur, 110
efficiency, definition of, 171
Eightfold Way, 213
Einstein, 40, 48, 84, 114, 116,
117, 118, 132, 133, 135, 139,
148, 150, 171, 187, 235, 242,
243
gravitational equation of,
243
electric dipole, 128, 192
electric monopole, 128
electromagnetism
electromagnetic force, 169,
186, 223, 226
electromagnetic force
strength, 199
electromagnetic interaction,
217, 225
electromagnetic wave theory of light, 130, 131, 150
electron magnetic moment, 193
electron neutrino, 203
electroweak force, 222, 234, 237,
265
electroweak symmetry, 239,
249
electroweak symmetry
breaking, 223
electroweak theory, 223
electroweak unification, 223,
226, 229
Elizabeth I, 80
emergent properties, 74, 271

Empedocles, 42
Encyclopédie (d'Holbach and
Thiry), 76
end of physics, 265
endothermic reaction, 170
energeticism, 111, 113
energy conservation, 192
energy levels, 152
Englert, François, 226
Enlightenment deism, 273
entropy, 100, 105, 108, 148
Epicurean atheism, 88
Epicureanism, 30, 31, 49, 56
Epicureans in Bible, 47
Epicurus, 11, 15, 21, 30, 31, 33,
35, 40, 41, 49, 50, 51, 56, 58,
242, 253, 261
Erasmus, 55
Eratosthenes of Cyrene, 34
European Center for Particle
Physics (CERN), 203
evangelical Christians, 176
evolution, 87, 262
evolution in atomism, 43
exchange force, 161
excited states of atom, 163
exothermic reaction, 170

Faraday, Michael, 124, 126, 128,
236
Faraday's law, 131
Fat Man (Nagasaki bomb), 175
Fermi, Enrico, 180, 181, 186, 190,
197, 200

Fermi model of beta decay, 200

Fermilab, 215, 227, 231, 232

fermions, definition of, 157

Feynman, Richard, 38, 185, 189, 190, 192

    Feynman diagram, 189, 190, 191, 192, 193, 200

field, definition of, 126

final cause, 43, 271

fine-structure constant, 199

fine structure of spectral lines, 154, 187

first law of thermodynamics, 96

Fitch, Val, 207

Fitzgerald-Lorentz contraction, 135

flatness problem, 244

fossil fuels, 175, 176, 182

frame of reference, definition of, 66

Franklin, Benjamin, 87, 220

French revolution, 87

Fukushima Daiichi nuclear plant, 178, 182

functional magnetic resonance imaging (fMRI), 232

fundamental forces, 186

Galilean invariance, 142

Galilean relativity, 66, 67, 69, 142

Galileo Galilei, 14, 56, 62, 63, 65, 70, 72, 73, 77, 87, 121, 132

    principle of relativity of, 132

    trial of, 64

gamma rays, 150, 169, 179

Gassendi, Pierre, 31, 56, 57, 59, 72, 120

gauge boson, 222

gauge symmetry or guage invariance, 211 219, 220, 226, 227

gauge transformation, 220

Gauss, Carl Friedrich, 105

    Gauss's law, 131

Gay-Lussac, Joseph Louis, 101

    Gay-Lussac's law, 101

Geber the Alchemist (Jäbir ibn Hayãn), 80

Geiger, Hans, 150

Gell-Mann, Murray, 197, 213, 214

general covariance, 142

general theory of relativity, 114, 117, 126, 142, 151, 152, 187, 242, 244, 253, 254, 265, 274

Genesis, book of, 242, 253

Georgi, Howard, 237

Germer, Lester, 154

Gibbs, Josiah Willard, 91, 104, 111, 272

Glashow, Sheldon, 222, 236, 237

global gauge transformation, 222

global warming, 176

gluon, 217, 224, 226, 229, 234

God particle, 227, 231

grand unification theories (GUTs), 236, 237, 244, 255
Grassie, William, 266
gravitational lensing, 247
gravitational repulsion, 251
Great Awakening, 87
Greenblatt, Stephen, 51, 54, 56
greenhouse effect, 176
ground state of atom, 163
Guralnik, Gerry, 226

hadron, 203, 204, 235
Hagen, Dick, 226
Halley, Edmund, 68
    Halley's comet, 69
Hanford, Washington, 175
Hargraves, Robert, 183
harmonic oscillator, 194
Hawking, Stephen, 253, 254
Hegel, Georg Wilhelm Friedrich, 88, 116
Heisenberg, Werner, 155, 160
    Heisenberg uncertainty principle, 160
Heisenberg-Born-Jordan matrix mechanics, 156, 158
heliocentrism, 87
Helm, George, 110, 112
Helmholtz, Hermann von, 96
Henry VIII, 55, 80
Herapath, John, 91
Herculaneum, 12, 30
Hero (or Heron) of Alexandria, 34

Hertz, Heinrich, 131
Higgs, Peter, 15, 226, 227
Higgs boson, 16, 158, 217, 223, 227, 228, 229, 231, 232, 234, 239, 249, 262, 265, 269
    Higgs decay, 229
    Higgs field, 228, 229, 251, 269
    Higgs mechanism, 223, 226, 228, 229, 232
higgsino, 249
Hindu religion, atoms in, 29
Hipparchus of Alexandria, 34
Hiroshima, 174, 175
Hitchens, Christopher, 261
Hobbes, Thomas, 31
Holden, Norman E., 83
holistic physics, 210
Holt, Jim, 38
Holy Roman Empire, 92
Honshu, 178
Hooke, Robert, 59, 68, 120, 122
horizon problem, 244
H-theorem, 105, 107
Hubble, Edwin, 243, 244
Hubble plot, 243, 250
Hubble Space Telescope, 251
Huygens, Christiaan, 34, 120–22
hyperfine splitting, 188
hyperon, 203, 205

ideal gas, 82, 101, 102, 271
Iliad (Homer), 29, 37
India, atoms in, 28

inflationary cosmology, 129, 256, 244

information theory, 107

Ingenhousz, Jan, 117

International Atomic Energy Agency, 177

International Space Station, 232

invariance, 140, 141, 263

Iphigena, 37

irreversibility, 109, 207
irreversibility paradox, 107, 109

Islam, golden age, 50

isotope, 85, 163

Jainism, atoms in, 29

Jefferson, Thomas, 47

Joule, James Prescott, 97

Justinian I (Roman emperor), 44

Kalām, 50, 51

Kamioka, Japan, 237, 238

Kanada, Hindu sage, 29

Kant, Immanuel, 89

Karaites (medieval Jewish sect), 50

Kepler, Johannes, 56, 67
Kepler's laws of planetary motion, 57, 68, 69
planetary system of, 153

Kibble, Tom, 226

kinetic theory of gases, 71, 100, 101, 104

Kirk, Geoffrey, 26

Klein, Felix, 112

K-mesons (*kaons*), 203

Koyrè, Alexander, 63

Krauss, Larry, 257, 258

Krumhansal, James, 232

Kuhn, Thomas, 95, 115

Lamb, Willis, 187
Lamb shift, 188, 189, 191, 192

La Mettrie, Julien Offray, 76

Laplace, Pierre-Simon, 272

Large Electron-Positron Collider (LEP), 232

Large Hadron Collider (LHC), 16, 210, 216, 218, 223, 225, 231, 239, 249

Lattice QCD, 235

Laurentian Library, Florence, 52

Lavoisier, Antoine, 83

law of conservation of mass, 84

law of definite proportions, 85

law of inertia, 69

law of large numbers, 107

law of multiple proportions, 85

Lawrence Berkeley Laboratory, 203

Lawrence Radiation Laboratory, 203

Lederman, Leon, 27, 227, 231, 233

Lee, Tsung Dao, 207

Lehninger, Albert L., 108

Leibniz, Gottfried Wilhelm, 77, 95

Lemaître, Georges-Henri, 242–44
lepton
    definition of, 203
    lepton number conserva-
        tion, 205, 206, 255
Le Sage, Georges-Louis, 91
Leucippus, 11, 23, 30, 38, 261
Lindberg, David, 72
linear vector algebra, 156
lines of force, 127
liquid fluoride thorium reactor
    (LFTR), 183
liquid-fuel reactor, 181
liquid-metal fast-breeder reac-
    tor, 181
Little Boy (Hiroshima bomb), 175
local gauge invariance, 222
Lock, John, 31, 59, 75
logical positivism, 114, 115
Lomonosov, Mikhail, 91
Lorentz invariance, 142
Loschmidt, Johann Josef, 103,
    104, 107, 109
    Loschmidt's number, 103
Lucretius, 12, 15, 31, 35, 37, 40,
    41, 52, 54, 55, 56, 117, 242, 261

Mach, Ernst, 14, 101, 110, 111,
    113, 114, 115, 116, 118
    Mach number, 114
    Mach's principle, 114
Machiavelli, Niccolò, 54
magnetic dipole, 129
    magnetic dipole moment, 158

magnetic moment
    of electron, 187, 188, 191
    of neutron, 188
    of the proton, 188
magnetic monopole, 129, 244,
    245
magnetic quantum number,
    153, 156, 157
magnetite, 206, 227
Maimonides, Moses, 50, 51
Man for All Seasons, 55
Manhattan Project, 158, 174,
    175, 186
Marsden, Ernest, 150
Martin, Richard, 183
matrix algebra, 155
matter-antimatter annihilation,
    172, 173
matter-antimatter asymmetry,
    255
matter-antimatter excess, 206
Maupertuis, Pierre Louis, 76
Maxwell, James Clerk, 101, 104,
    105, 107, 119, 130, 131, 132,
    236, 272
    Maxwell's demon, 107
    Maxwell's equations, 130,
        132, 133, 222
Maxwell-Boltzmann distribu-
    tion, 101, 105
Mayer, Julius Robert von, 9
mechanical philosophy, 72
Medici family, 52
Mendeleev, Dmitri, 84, 85

meson
   definition of, 204
   Yukawa theory of, 197
metallurgy, 83
Michelangelo, 52
Michelson, Albert, 132
Michelson and Morley experiment, 132
Middle Ages, 22, 24, 49, 62, 76,
   79
Milky Way, 247
Millikan, Robert A., 149
minimal SU(5), 237
mirror symmetry, 141, 142, 206
molten-salt reactor, 181
momentum conservation, 192
monopole problem, 244
More, Thomas, 55, 56
Morley, Edward, 132
Muhammed, 50
multiverse, 48, 256, 262, 274
muon, 186
   discovery of, 198
   muon neutrino, 203
mutually assured destruction
   (MAD), 175

Nafe, John E., 188
Nagasaki, 174, 175
Napoleon, 93
National Nuclear Data Center,
   83
Nausiphanes, 30
Neddermeyer, Seth, 186

Ne'eman, Yuval, 213
Nelson, Edward B., 188
Neurath, Otto, 114
neutralino, 249
neutrino
   discovery of, 186
   neutrino mass, 216
   neutrino oscillation, 217
neutron
   discovery of, 151
   neutron beta decay, 205
New Age, 210, 270
Newton, Isaac, 14, 59, 61, 62, 68,
   70, 72, 73, 81, 88, 120, 121, 122,
   139, 142, 236
   corpuscular theory of light,
      150
   law of gravity, 126, 142, 247
   laws of motion, 69, 82, 139,
      142
   theory of colors, 121
   theory of light, 120
Newtonian mechanics, 65, 73,
   91, 100, 109, 133, 272
Newtonian world machine, 272,
   273
Niccoli, Niccolò, 52
Nicholas of Autrecourt, 49
Nixon administration, 181
noble elements, 165
Noether, Emmy, 142
   Noether's theorem, 142, 143
Nolan, Lawrence, 73
non-Euclidean geometries, 244

nuclear bomb, 185
nuclear democracy, 208
nuclear fission, 174
nuclear fusion, 172, 173
nuclear proliferation, 175
nucleon number, 162
    nucleon number conservation, 168

O(3) symmetry group, 213
Oak Ridge, Tennessee, 175, 185
Ockham's razor, 49, 274
Odyssey (Homer), 29
omega-minus baryon, 214
Oppenheimer, Robert, 167
orbital angular momentum
    quantum number, 153, 156,
    164
Ostwald, Wilhelm, 110, 112, 115,
    118
O(3) symmetry group, 213
Oxford calculators, 63

pair annihilation, 190
pair production, 190
parity, 207
    parity symmetry (mirror
    symmetry), 207
    parity symmetry violation,
    207
partial differential equations,
    155
Pascal, Blaise, 105
    Pascal's Wager, 105

Pauli, Wolfgang, 155, 157, 186
    Pauli exclusion principle,
    157, 163, 164, 224
Penrose, Roger, 253, 254
periodic table, 15, 84, 85, 157,
    164, 165, 168, 213, 266
perpetual-motion machine, 95
Perrin, Jean Baptiste, 40, 118
philosopher's stone, 79, 81
phlogiston, 83
photoelectric effect, 147, 148,
    149
photon theory of light, 148, 149
Physical Review Letters (journal),
    226
Physics Letters (journal), 234
pilot wave, 159
pi meson (pion), discovery of,
    198
pion. *See* pi meson (pion), discovery of
pion-exchange model, 198, 225
Planck, 104, 118, 147, 148, 149,
    194
    Planck length, 253, 254
    Planck quanta, 115
    Planck time, 253, 254
    Planck's constant, 148, 150
Plato, 13, 22, 47, 48
Plotinus of Lykopolis, 45
Pocono Manor Inn, 189
Pompeii, 12
Pontifical Academy, 242
Pope John XXII, 80

Pope John XXIII (the "antipope"), 52
Pope Pius XII, 242
Popper, Karl, 115
positivism, 14, 113
positrons, discovery of, 158
pressurized-water reactor, 180
primary properties, 73
Princeton University, 189
Principia (Newton), 69, 73
principle of Galilean relativity, 132
principle quantum number, 156,
163, 164
proton decay, 237, 255
Proust, Joseph, 85
Ptolemaic model, 67
Ptolemy, Claudius, 34, 64, 120
Pullman, Bernard, 29, 48, 51, 88,
267
Putnam, Hilary, 115
Pyle, Andrew, 21

qualia, 75
quanta, 148
quantum chromodynamics
(QCD), 223, 224, 225, 250
quantum electrodynamics
(QED), 158, 189, 193, 194, 197
quantum field theory, 158, 195,
208, 210, 222, 268
quantum fluctuations, 192
quantum gravity, 254
quantum of action, 152
quantum spirituality, 210

quantum tunneling, 173, 255, 274
quark model, 214
Quine, Willard Van Orman, 115
quintessence, 123

Rabi, Isidor, 188
radioactivity, discovery of, 150
Raman, Varadaraja V., 266
Raven, John, 26
Reagan, Ronald, 231
reductionism, 17, 241
reference frame, definition of,
140
Reformation, 15, 62
refraction, 122
Reichenbach, Hans, 114
relativistic quantum field theory, 188, 194, 197
relativistic quantum theory, 187
Renaissance, 14, 15, 22, 31, 51,
62, 262
renormalization, 189, 229
rest energy, definition of, 139
Retherford, Robert, 187
Romantic movement, 87
Röntgen rays, 115
Rosenberg, Alex, 22
Royal Institution, 84
Royal Society, 92, 93, 100, 120,
122
Rubbia, Carlo, 223
Rutherford, Ernest, 150
model of the atom of, 151,
152

Rutherford-Geiger-Marsden experiment, 215

St. Louis World's Fair, 116
Sakharov, Andrei, 255
Salam, Abdus, 222, 223, 229, 236
Sample, Ian, 223, 231
Savonarola, Girolamo, 54
scalar field, definition of, 126
Scanning Tunneling Microscope, 14
Schofield, Malcolm, 26
scholasticism, 44
Schopenhauer, Arthur, 88, 89, 116
Schrödinger, Erwin, 155, 159
    quantum mechanics of, 158
    Schrödinger equation, 155, 156, 165
    Schrödinger wave mechanics, 156, 159
Schwinger, Julian, 189, 193, 194
Scientific American (journal), 258
secondary properties, 73
secondary qualities, 74, 75
Second Coming of Christ, 82
second law of thermodynamics, 95, 100, 105, 107, 110, 147
Sedley, David, 25
Shannon, Claude, 107
singularity, 253
S-Matrix Theory, 210
Socrates, 25
Sommerfeld, Arnold, 113, 153

sound waves, 132
space-rotation invariance, 141
space-rotation symmetry, 141, 143
space-time diagrams, 190
space-translation invariance, 141
space-translation symmetry, 141, 143
special theory of relativity, 117, 132, 148, 153, 158
spherical symmetry, 141
spin
    definition of, 157
    spin quantum number, 157, 187
spontaneous symmetry breaking, 226, 227
Stallings, A. E., 37
Standard International System of units, 98, 99
standard model of cosmology, 248
standard model of elementary particles and forces, 15, 159, 195, 204, 208, 210, 211, 216, 218, 222, 225, 226, 229, 231, 234, 236, 237, 246, 248, 257, 264, 265, 268, 270
state vector, 156, 220
statistical mechanics, 104
sterile neutrinos, 249
stoicism, 45
Stoics, 44, 48

Stoney, George, 118
strangeness, 203
     strangeness conservation,
         203, 205
Strato of Lampsacus, 34
Strauss, Lewis, 177
string theory, 17
strong force strength, 200, 225
strong nuclear force, 169, 172,
     186, 203, 205, 217, 222–24, 236,
     265
Stückelberg, Ernst, 190, 192
Superconducting Super Col-
     lider (SSC), 231, 223
superconductivity, 232
Super-Kamiokande, 217, 238
supersymmetry (SUSY), 238,
     239, 249
SU(2)´U(1) symmetry group,
     236
SU(3) symmetry group, 213
swerve, 33, 41, 45, 51, 272
symmetry
     symmetry breaking, 234
     symmetry groups, 236
     symmetry principles, 141

Tao of Physics, The (Capra), 210
tauon, 203
     tauon neutrino, 203
Tevatron, 232
Thales of Miletus, 47
theory of everything, 239, 265
therapeutic touch, 263

thermodynamics. See first law
     of thermodynamics; second
     law of thermodynamics
Thierry of Chartres, 49
Thompson, Benjamin (Count
     Rumford), 92
Thomson, J. J., 118
Thomson, William (Lord
     Kelvin), 99, 107
't Hooft, Gerhardus, 229
thorium, 179, 181, 182
thorium fluoride reactor
     (LFTR), 181
Three Mile Island, 177, 178,
     182
time dilation, 134
time-reversal symmetry viola-
     tion, 207, 208
time-translation invariance, 141
time-translation symmetry, 141,
     143
Tomonaga, Sin-Itiro, 189
top-down causality, 271
Torricelli, Evangelista, 16
Tower of Pisa, 65
twin paradox, 135
Type 1a supernova, 250

Ukraine, 177
ultraviolet catastrophe, 146
uncertainty principle, 161, 192,
     194, 253, 272
unification, 236
unitarity, definition of, 221

United Nations Development Agency, 177
University of California at Berkeley, 208
University of California at Los Angeles (UCLA), 203
University of Hawaii, 203, 238
University of Texas, 232
urban air pollution, 176
Utopia (More), 55

vacuum polarization, 192
valence electrons, 165
Van der Meer, Simon, 223
Vatican Library, 54
vector fields, 126
vector potential, 220, 222
Veltman, Martinus, 229
Vesuvius, 12, 30
Vienna Circle, 114
Vienna Philosophical Society, 116
Viennese Academy of Sciences, 114
Virgil, 52
virtual particles, 192
vis viva, 95, 97
von Helmholtz, Hermann, 96
von Mayer, Julius Robert, 95

Waterston, John James, 100, 101
wave function, 155, 156, 159, 160, 220
   of the void, 258

wave-particle duality, 150, 154
wave theory of light, 121, 122, 146
Waxahachie, Texas, 231
weak boson, 201, 234
weakly interacting massive particles (WIMPS), 249
weak nuclear force, 169, 172, 187, 200, 205, 217, 222, 223, 226, 229, 234, 236, 265
Weinberg, Alvin, 180, 181
Weinberg, Steven, 213, 222, 223, 229, 231, 236, 252
White, Michael, 81, 121
Whore of Babylon, 81
Wigner, Eugene, 181
Wilczek, Frank, 258
Wilkinson Microwave Anisotropy Probe (WMAP), 245–47
William of Conches, 49
William of Ockham, 49
Willis, Thomas, 59
WIMP (weakly interacting massive particles), 249
WMAP (Wilkinson Microwave Anisotropy Probe), 245–47
World Health Organization, 176, 177
World War II, 158, 174, 175, 185, 188
Wren, Christopher, 68
Wu, Chen-Shiung, 207

Yang, Chen Ning, 207
Yang-Mills theories, 229

Yergin, Daniel, 178
Young, Thomas, 123
Yukawa, Hideki, 197, 198

Zeeman effect, 153, 154
Zeno of Cittium, 44
zero-point energy, 194, 252
Zweig, George, 213, 214